Water Sustainability in the Pulp & Paper Industry 2011

Providence, Rhode Island, USA
10 – 11 November 2011

ISBN: 978-1-5108-1776-0

Printed from e-media with permission by:

Curran Associates, Inc.
57 Morehouse Lane
Red Hook, NY 12571

Some format issues inherent in the e-media version may also appear in this print version.

Printed by Curran Associates, Inc. (2016)

For permission requests, please contact AIChE
at the address below.

AIChE
120 Wall Street, FL 23
New York, NY 10005-4020

Phone: (800) 242-4363
Fax: (203) 775-5177

www.aiche.org

Additional copies of this publication are available from:

Curran Associates, Inc.
57 Morehouse Lane
Red Hook, NY 12571 USA
Phone: 845-758-0400
Fax: 845-758-2634
Email: curran@proceedings.com
Web: www.proceedings.com

TABLE OF CONTENTS

WATER SUSTAINABILITY
in the Pulp & Paper Industry

Co-Produced by

AIChE TAPPI

The Providence Biltmore Hotel

Providence, RI

November 10-11, 2011

Website:

http://water.aiche.org/conferences

Workshop Overview

Industry Background:

The pulp and paper industry is among one the largest manufacturing sectors in 48 states. It is also one of the largest water users. However, nearly 88% of this water is treated and returned to the environment, greatly reducing the amount of consumed water. Nevertheless, there are expanding efforts to overcome water use challenges. These efforts include reducing the use of fresh water in the pulping and papermaking processes, developing technologies that treat and reuse process water, and developing closed-loop systems.

This initial event begins to address some of these challenges to help provide a better understanding of water issues in the pulp and paper sector as well as define water sustainability initiatives in the industry.

Workshop Goal:

This is a 2 day intensive workshop features invited presentations focusing on water sustainability issues in the pulp and paper industry. The objective is to develop a more detailed best practices effort as the pulp and paper industry moves forward in reducing their water footprint and investing in expanding their water sustainability initiatives.

This Program Features 4 Topical Sessions:

- External Pressures and Drivers
  - Global Competitiveness – countries with more lax environmental regulations
  - GHG Footprint – pressures from buyers like Walmart to reduce environmental footprint
  - Regulations on water use, contaminant levels

- Water Use Challenges
  - Reduce use of fresh water in the pulping and papermaking processes
  - Develop technologies that treat and reuse process water
  - Develop closed-loop systems

- New Technology Needs
  - Separation processes – membrane technologies, fast-acting systems, process modeling
  - Contaminant removal – trace metals in process waters, pulp fibers
  - Low-grade heat recovery from process water – separation out of contaminants, new technologies
  - Waste water treatment – technological advances

- Barriers to Development
  - Funding for R&D, pilot plant studies
  - Collaboration with membrane community
  - Greater understanding of process chemical transport and other phenomena

Workshop Agenda

Thursday, November 10, 2011

8:00 – 8:30am	Continental Breakfast
8:30 – 10:00am	**Session I – External Pressures and Drivers, Panel Discussion** Welcome (20-30min) ➢ Introductions of Organizing Committee and Panel Members ➢ Workshop Goals and Overview ➢ Panel Discussion with: **Jerry Schwartz,** *Senior Director of Environmental Policy*, AF&PA **Nicole Tanner**, *Assistant Global Water Roundtable Coordinator, AWS* **Laura Thompson**, *Dir. of Tech. Mktg & Sustain Development*, SAPPI
10:00 – 10:30am	Break
10:30 – 12:00pm	Continue Session I's Q&A
12:00 – 1:30pm	Lunch
1:30 – 3:00pm	**Session II – Water Use Challenges** ➢ Presentations by: **Barry Malmberg**, *Sr. Research Engineer,* NCASI **Ron Brown***, Executive Director*, Agenda2020 Technology Alliance
3:00 – 3:30pm	Break
3:30 – 5:00pm	Wrap up Session II
5:00 – 7:00pm	Reception (Optional)

Friday, November 11, 2011

8:00 – 8:30am	Continental Breakfast
8:30 – 10:00am	**Session III – New Technology Needs** ➢ Presentations by: **Steven Hatch**, *Global Technology Leader,* Nalco Company **Vincent Caprio**, *Chief Operating Officer*, Water Innovations Alliance **Dr. Mrinal Mahapatro**, *Staff Scientist*, Pall Corporation
10:00 – 10:30am	Break
10:30 – 12:00pm	**Session IV – Barriers to Development**
12:00 – 1:30pm	Lunch
1:30 – 3:00pm	Wrap-up & closing comments

Featured Speakers

G. Ronald (Ron) Brown, PhD

Executive Director, Agenda 2020 Technology Alliance

Ron Brown is Executive Director of the Agenda 2020 Technology Alliance, an industry-led partnership with government and academia that identifies technology needs of the forest products industry and facilitates collaborative R&D programs to address these needs. Agenda 2020 is housed within the American Forest & Paper Association, in Washington, DC. Ron held leadership positions in industrial research, engineering, and manufacturing with MeadWestvaco Corporation spanning more than 30 years. Since joining Agenda 2020 in 2008, he has concentrated on transformational technologies for the forest products industry, with emphasis on reducing energy intensity, greenhouse gas emissions, and water intake in mills.

Vincent Caprio

Chief Operating Officer, Water Innovations Alliance

Mr. Caprio is one of the foremost advocates for government funding of emerging technologies at both the State and Federal levels. Mr. Caprio has testified before the state legislatures of New York and Connecticut, and has participated in the NNI Workshop. Mr. Caprio is the founder of the Water Innovations Alliance Conferences and has successfully produced 3 Water Innovations Alliance Conferences. He is the Event Director of the 2nd Annual Water Innovations Alliance Conference, September 27-29, 2010 in Chicago, IL. In addition, Mr. Caprio has appeared on numerous Connecticut TV and radio stations discussing the impact of Water and Nanotechnology.

Steve Hatch, PhD

Global Technology Leader, Nalco Company

Steven Hatch received a B.S. in Chemistry from Brigham Young University in 1985 and a Ph. D. in Physical Chemistry from University of Texas at Austin in 1990 where he studied surface photochemistry and the Surface Raman effect with Professor Alan Campion. In 1990 he received a Post-Doctoral Fellowship with Exxon Corporate Research and Engineering to investigate interfacial chemistry, in-situ, utilizing nonlinear spectroscopies. Steve began work at Nalco Chemical Co. in 1992 in the Research Analytical labs and later managed the Cooling Water Research department. He currently serves as a Global Technology Leader for the Wastewater, Membranes and Process Solutions; Oil Sands and Produced Water; and Research Analytical areas. Steve has been closely involved in development of Nalco's 3DTRASAR technology and was part of the team recognized with the United States EPA Presidential Green Chemistry Award for this advancement in 2008. Steve has authored 18 publications and holds 6 US patents.

Dr. Mrinal Mahapatro

Staff Scientist, Pall Corporation

Dr. Mrinal Mahapatro is a Staff Scientist in the Scientific & Laboratory Services (SLS) Group at Pall Corporation. He has been at this position for the past three years. Prior to Pall, Dr. Mahapatro was a Senior Chemist in the Aerospace division of PPG Industries in California developing coatings for aircraft transparencies. In addition, he carried out several Post-Doctoral fellowships in the general area of Surface Chemistry / Nanotechnology. Dr. Mahapatro holds a Doctorate Degree in Chemistry.

Barry Malmberg, PhD
Sr. Research Engineer, NCASI

Barry Malmberg is a Senior Research Engineer at the National Council for Air and Stream Improvement. He has been at NCASI for six years working in the areas of water and energy sustainability for the pulp and paper industry. Barry has a Ph.D. in Chemical Engineering from the University of Idaho.

Jerry Schwartz
Senior Director of Energy and Environmental Policy, AF&PA

Jerry Schwartz is the Senior Director of Energy and Environmental Policy at the American Forest & Paper Association. In that position, he helps establish and implement AF&PA's energy and environmental policy and advocacy. Jerry also is the Sustainability Team Leader, directing AF&PA's work on various sustainability initiatives, including the *Better Practices, Better Planet 2020* sustainability goals program. Before joining AF&PA, he owned a litigation support consulting firm; traveled around the world for a year with his wife; and worked in the Environmental Enforcement Section of the Department of Justice and at EPA.

Nicole Tanner
Assistant Global Water Roundtable Coordinator, AWS

Nicole Tanner has been working with WWF-US's Market Transformation Unit on commodity and freshwater issues since 2008. In 2009, she also began work with the Alliance for Water Stewardship (AWS), on behalf of WWF, as the Assistant Global Coordinator of the Water Roundtable process. Nicole has degrees in Political Science and International Relations from Elmira College (NY) and a Masters in International Development from the University of Denver's Korbel School of International Studies. Originally from upstate New York, she currently lives in Washington, DC and is learning to balance city life with a love of all things outdoors. Through her work with WWF and AWS, Nicole continues to advocate for sustainable development and freshwater stewardship globally.

Laura M. Thompson, PhD
Director of Technical Marketing and Sustainable Development, SAPPI

Laura Thompson is a recognized leader and fact-based voice in the world of paper and sustainability, building her expertise over the past 15 years in the paper industry. Dr. Thompson has a wide breadth of experience in the industry, including R&D, mill environmental, product development for specialties and coated fine paper, and most recently, sustainability.

Dr. Thompson joined Sappi Fine Paper North America in 2006 and serves as Director of Sustainable Development, providing thought leadership and setting strategy for the company's sustainability goals and the corporate governance around those efforts. In addition, Dr. Thompson is at the helm of Sappi's eQ brand which includes the eQ Journal, eQ Microsite (including informative videos, white papers, and Sappi's original eQ Tool which shows paper buyers the environmental impact of their purchases), and is the author of The Environmental Quotient blog which is hosted by Sappi. The Environmental Quotient is a platform for the entire industry to participate in an online discussion around a myriad of topics related to sustainability. Dr. Thompson is dedicated to offering immediacy, transparency, and candidness -- all supported by a solid foundation in fact and science. The Environmental Quotient blog can be accessed by visiting www.eq.tumblr.com.

Dr. Thompson earned a B.S. in Chemical Engineering from the University of New Hampshire and an M.S. and PhD in Paper Science from the Institute of Paper Science and Technology. She is active with several industry associations, including The Association for Magazine Media Environmental Working Committee; Green Blue's Advisory Board for The Paper Life Cycle and Forest Products Working Group; TwoSides US Sustainability Committee; TAPPI Sustainability Forum 2012 Program Committee; and, Dr. Thompson served as a Judge for Net Impact Sustainability Challenge on Scorecard Development.

Water Sustainability in the Pulp & Paper Industry

The Providence Biltmore Hotel, Providence RI
November 10-11, 2011

Organizing Committee: Eric Bober, Nexant; Katherine Broadus, Nalco, An Ecolab Company

Workshop Overview

Industry Background:

The pulp and paper industry is among one the largest manufacturing sectors in 48 states. It is also one of the largest water users. However, nearly 88% of this water is treated and returned to the environment, greatly reducing the amount of consumed water. Nevertheless, there are expanding efforts to overcome water use challenges. These efforts include reducing the use of fresh water in the pulping and papermaking processes, developing technologies that treat and reuse process water, and developing closed-loop systems.

This initial event began to address some of these challenges and provided a better understanding of water issues in the pulp and paper sector as well as defined water sustainability initiatives in the industry.

Workshop Goal:

This was a 2 day intensive workshop featuring invited presentations focusing on water sustainability issues in the pulp and paper industry. The objective was to develop a more detailed best practices effort as the pulp and paper industry moves forward in reducing their water footprint and invests in expanding their water sustainability initiatives.

This Program Featured 4 Topical Sessions:

- External Pressures and Drivers
  - Global Competitiveness – countries with more lax environmental regulations
  - GHG Footprint – pressures from buyers like Walmart to reduce environmental footprint
  - Regulations on water use, contaminant levels
- Water Use Challenges
  - Reduce use of fresh water in the pulping and papermaking processes
  - Develop technologies that treat and reuse process water
  - Develop closed-loop systems
- New Technology Needs
  - Separation processes – membrane technologies, fast-acting systems, process modeling
  - Contaminant removal – trace metals in process waters, pulp fibers
  - Low-grade heat recovery from process water – separation out of contaminants, new technologies
  - Waste water treatment – technological advances
- Barriers to Development
  - Funding for R&D, pilot plant studies
  - Collaboration with membrane community
  - Greater understanding of process chemical transport and other phenomena

Summary - Session I: External Pressures and Drivers

Water is a much discussed subject in the social media as water consumption is becoming an increased concern in the public eye. Studies have shown that 90% out of 32,000 people in 15 countries deem water pollution a serious problem. The Pulp and Paper Industries is one of the top manufacturing sectors in the U.S. and currently ranks as the 4th largest water user. Many industries, government, and non-governmental agencies are taking action to address these concerns.

Following the presentations by Jerry Schwartz (AF&PA), Nicole Tanner (WWF/AWS), and Laura Thompson (SAPPI), a panel and floor discussion took place to determine the external pressures and drivers to water challenges, list the priorities, and suggest approaches to address these issues in the pulp and paper sector. Some of these drivers discussed include:

1. **Regulatory Drivers**
   Regulations on water use could drive people to use less water. A specific example is the EPA Cooling Water Intake Structure (CWIS) Rule. This rule places a threshold on water use for companies that use below 2 million gallons per day. Below this line, there is no "large" environmental impact. The Great Lakes Regional Compact is another example of a local project that places certain limitations on businesses depending on the amount of water withdrawal.

2. **Marketplace Drivers**
   There is an initiative by the Alliance for Water Stewardship (AWS) to develop a global water stewardship program that "recognizes and rewards water users and managers who take significant steps to minimize their water use and impact." The AWS also created a global water roundtable that is making efforts to develop a water sustainability standard by July 2013. This serves as the basis for water certification.

   Customer goals in the marketplace are also a huge driver. It was agreed that market perception is an essential part of getting companies to establish sustainability goals. The AF&PA is the only organization representing US manufacturers in any industry with such an extensive suite of quantifiable and verifiable sustainability goals. As an example, AF&PA was able to exceed goal expectations in reducing Greenhouse Gas emissions to 20% by 2012. AF&PA then moved to higher expectations, bumping the numeric goal to 15% more by 2020. Reviewing performance on a regular basis will greatly contribute to setting sensible numeric goals that make sense.

3. **Climate Events (e.g. Drought)**
   There are always risks of climate variability. Mills need to ask themselves: are we in a sustainable area? What is the procedure during an occurrence?

4. **SEC Guidance on Water for Disclosure (and for carbon)**

5. **Water Risks (e.g. physical, regulatory, environmental, financial, and reputational)**

6. **Energy Conservation**
   Water technologies are driven by energy projects. Water treatment processes and management come at an energy cost and may have a detrimental effect on energy conservation.

7. **Competing Demands on Water**
   Businesses are acquiring permits and licenses for watershed privileges. Pulp and Paper industries are also situated where there is ample water supply. Pulp and paper mills pay for the cost of heating, pumping, and treating water but the overall cost of water is not well understood.

In the future, stakeholders with economic backgrounds may put a price on water to reduce water consumption.

8. **Stakeholder Concerns**
   Companies respond to competition. It is essential that companies clearly define the scope and what they are comparing to when reporting data. These include questions such as what are the metrics? Why does it matter? How do we measure the water balances? Is it total water? If 88% of total water intake gets returned, what about the water in the wood? Are the numbers tied to sustainability when applying water footprint networking?

The next questions addressed the actions and responses needed to overcome these challenges, pressures, and drivers as previously discussed: The group agreed that a **clear set of definitions and metrics are needed**. Although a multitude of organizations are attempting to define these metrics, the problem arises when we end up with >50 different set of metrics. In this case, **what are the right metrics?** In Life Cycle Assessment (LCA), measurements are heavily dependent on the scope you are studying. Allocation of methods in LCA can also influence results, potentially driving false conclusions and bad decisions. Therefore, there is a strong need to **determine the key leading groups and come up with a standard**. In response, AIChE is currently making strong efforts to come up with a standard set of ~8 key metrics that will give about 80% accuracy. Awareness has even reached the Dow Jones Index, which has been revised to include a water component.

At the WWF, water efforts are working together to form a standard in preserving water sheds. This includes the CEO water mandate (CEOwatermandate.org). The first draft is expected to be out for review in March 2012.

Finally, we need to **look at the list of water projects and find common metrics**. This is one of the pilots for the pulp and paper industry.

Summary – Session II: Water Use Challenges

Session II opened with presentations from Barry Malmberg (NCASI) and Ron Brown (Agenda2020 Technology Alliance). The **takeaways and remaining water challenges in the pulp and paper industry** from this session include:

1. Lack of external driving forces (e.g. economic or regulatory)
2. Increased process control demands
3. Unknown water quality tolerances in operations
4. Too many complicated and undefined metrics
5. Standard definitions (water use – consumptive; discharge; process water, or process and cooling water)
6. Challenge to convey global message because localized issue
7. Difficult to convey perception of water use – what's good and what's bad
8. Putting the message into the watershed context
9. Habits (water has historically been plentiful)
10. Bias of machine operators against reuse due to past bad experience
11. Breakthrough for felt cleaning in press shower; consistent quality effluent for recycle use
12. Can the goals be met with existing vs. new technologies?
13. Variability in the age of equipment PM or pulp mill

Summary – Session III: New Technology Needs

Man-made and natural causes have been a strong driving factor to lower water consumption and water use around the world. In response, industries are striving to research and develop new technologies that will facilitate mill operations and help achieve these goals. Steven Hatch (Nalco), Vincent Caprio (Water Innovations Alliance), and Dr. Mrinal Mahapatro (Pall Corporation) discussed the developing technologies in the market today and the future technologies needed in water treatment processes.

The new technology needs include:

- Development of non-contact sensors that are antifouling
- Removal of sticky fibers from membranes for separation
- Potential applications for targeting water reuse.
- Water specifications for end-use applications
    - Examples: mechanical seals cooling, machine showers, nozzles
    - How much contaminant can a modern press felt take?
- Pulp washing at higher consistency
    - There is a need to improve mixing performance in both mechanical and non-mechanical mixers
    - Define the range/benchmark for washers
    - What are the specifications and summaries of the best washers that are available?
- Challenges in treatment of filtrate from initial Kraft bleaching stages
    - Low ph operations challenges and hot ph on peroxides
    - Filtration in corrosive environments
- Monitoring and controlling Save-All clear filtrate quality in order to reuse water
    - Challenges in identifying the right probes/sensors for that application
    - Need technologies to control dissolved and suspended solid levels
    - Some commercial sensors could cost around $6k. Therefore, there is a need for antifouling sensors to keep costs low.
- New treatment and cleaning chemistries to improve throughput of membranes in waste treatment
    - What level of removal is required to make substantial improvement?
    - Again, water specifications are needed
- New technologies for tertiary treatment of waste effluent to supply water with acceptable quality for reuse ("treat to need")
    - Capital expenses are high so there is a need to improving treatment costs
    - Tertiary treatment required due to low EPA levels
- Examine total costs optimizations (energy, water, and environmental costs)

Summary – Session IV: Barriers to Development

For the closing session, we opened the floor to discussion regarding all barriers to development of these new technologies.

It was agreed that the main barrier is **defining the specifications**. What are the specifications on water contaminants? What is the level we need to reach? What are the water operations specifications? How are mills reusing water? What types of failure mechanisms exist in water operations? **Conducting surveys** of the mills would be an initial approach to answering these questions.

Furthermore, **site testing is also needed for tertiary clean up technologies**. But, where can these tests be done? And how do we keep the costs down? There are barriers to advanced improvements since capital expense is a major issue and a great amount of equipment is involved.

So how do we overcome these barriers? The best solution is to reach **out to the "desperate groups" who have no fresh water available** (e.g. Israel, Singapore, etc.) and find out what they are doing. How are they setting and meeting their goals? By focusing on water-deprived regions, we can explore the options that exist.

American Forest & Paper Association

www.afandpa.org

Water Sustainability for the Pulp and Paper Industry

WATER SUSTAINABILITY in the Pulp & Paper Industry

November 10-11, 2011 • Providence Biltmore Hotel • Providence, RI

Co-Produced By AIChE TAPPI

External Pressures and Drivers

Jerry Schwartz

November 10, 2011

BETTER PRACTICES BETTER PLANET 2020
Continuing AF&PA's Commitment to Sustainability

Overview

- Why Focus on the Forest Products?
- Specific Drivers
  - Regulatory
  - Marketplace
  - Drought
- Industry Response

BETTER PRACTICES BETTER PLANET 2020
Continuing AF&PA's Commitment to Sustainability

2

Why Focus on Forest Products?

- Water Withdrawals: Industry is the Fourth Largest Category Behind Electric Power, Agriculture, and Public Supply

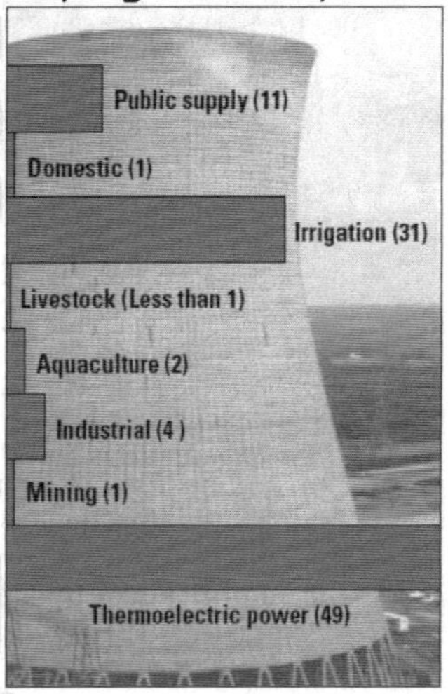

"Self-supplied industrial withdrawals were an estimated 18,200 Mgal/d, about 4 percent of total withdrawals. Industrial water use includes water used in manufacturing and producing commodities, such as food, paper, chemicals, refined petroleum, wood products, and primary metals."

- *Summary of Estimated Water Use in the United States in 2005, US Geological Survey*

BETTER PRACTICES BETTER PLANET 2020
Continuing AF&PA's Commitment to Sustainability

3

Why Focus on Forest Products?

- Large Industrial Water User

BETTER PRACTICES BETTER PLANET 2020
Continuing AF&PA's Commitment to Sustainability

4

Regulatory Drivers

- Highly Regulated
  - Technology-Based and Water Quality- Based
  - TSS, BOD, AOX, Toxicity, TMDLs, *etc.*
- Shift to Water Quantity

BETTER PRACTICES BETTER PLANET 2020
Continuing AF&PA's Commitment to Sustainability

Regulatory Drivers: EPA Cooling Water Intake Structures (CWIS) Rule (CWA Section 316(b))

- Requires the Design, Construction and Capacity of CWIS Reflect the Best Technology Available for Minimizing Adverse Environmental Impact
- Proposed Existing Facilities Rule: April 2011
- Final Rule: July 2012
- Separate Standards for Impingement Mortality and Entrainment
- Uniform Numeric Impingement Standards for all Facilities Over 2 Million Gallons/Day (MGD) or Reduction of Intake Velocity to 0.5 feet/second
- Promotes (but does not require) Closed-Cycle Cooling
- "Reward" Water Reuse?

BETTER PRACTICES BETTER PLANET 2020
Continuing AF&PA's Commitment to Sustainability

6

Regulatory Drivers: EPA Cooling Water Intake Structures Rule

BETTER PRACTICES BETTER PLANET 2020
Continuing AF&PA's Commitment to Sustainability

7

Great Lakes – St. Lawrence River Basin Water Resources Compact

- ALL Withdrawals > 100,000 Gallons/Day Must Be "Registered"
- Permits Required For **All New Or Increased** Withdrawals Above A Defined (By Each State) Threshold (Varies From 100,000 Gallons/Day To Millions)
- Permits Must Meet "Decision Making Standard"
  - "No Significant Cumulative or Individual Adverse Impact to the Quantity or Quality of the Waters or Water Dependent Natural Resources and the Applicable Source Watershed"
- New Conservation Requirements For ALL Users – Existing Or New
- Approval Of >5 Mgd Consumptive Uses Must Be Reviewed By The Compact Regional Body

BETTER PRACTICES BETTER PLANET 2020
Continuing AF&PA's Commitment to Sustainability

Marketplace Drivers: Water Initiatives

- Aquawareness
- Alliance for Water Stewardship/Global Water Roundtable
- BIER Water Footprint Working Group
- Carbon Disclosure Project -Water Disclosure
- Ceres Aqua Gage
- Corporate Water Gauge
- Federation House Commitment to Water Efficiency
- Global Environmental Management Initiative (GEMI)
- Global Reporting Initiative
- Global Water Tool (WBCSD) UNEP/SETAC Life Cycle Initiative
- Strategic Water Management Framework (Australia minerals)
- UN CEO Water Mandate
- Water Brief for Business
- Water Footprint Network
- Water Footprint Neutrality and Efficiency Project (UN)
- Water Neutral Offset Calculator
- WaterSense Certification Scheme
- ISO Water Footprint Standard

BETTER PRACTICES BETTER PLANET 2020
Continuing AF&PA's Commitment to Sustainability

Source: NCASI

Marketplace Drivers: Water Initiatives

Water Footprinting	Water Sustainability/ Stewardship	Disclosure/ Risk Management
Water Footprint Network	Alliance for Water Stewardship/Global Water Roundtable	Carbon Disclosure Project: Water Disclosure

BETTER PRACTICES BETTER PLANET 2020
Continuing AF&PA's Commitment to Sustainability

10

Water Footprint Network

- •Shows how water footprints can be calculated for individual processes and products, as well as for consumers, nations and businesses
  •Contains detailed worked examples of how to calculate green, blue and grey water footprints
  •Describes how to assess the sustainability of the aggregated water footprint within a river basin or the water footprint of a specific product
  •Includes an extensive library of possible measures that can contribute to water footprint reduction

BETTER PRACTICES BETTER PLANET 2020
Continuing AF&PA's Commitment to Sustainability

11

Marketplace Drivers: Alliance for Water Stewardship

ALLIANCE FOR WATER STEWARDSHIP

- The AWS, working with engaged stakeholders from around the world, is developing a global water stewardship program and a permanent organization to house this initiative. We need to create ***a program that recognizes and rewards water users and managers who take significant steps to minimize their water use and impacts.*** At the heart of the program is the development of an International Water Stewardship Standard (referred to herein as "the Standard") based on critical aspects of water stewardship such as managing flow, water quality, habitats and ensuring good governance.

BETTER PRACTICES BETTER PLANET 2020
Continuing AF&PA's Commitment to Sustainability

12

ALLIANCE FOR WATER STEWARDSHIP

Who is the Alliance for Water Stewardship?

The Nature Conservancy
Protecting nature. Preserving life.
nature.org

WWF

Water Environment Federation
the water quality people

PACIFIC INSTITUTE

IWMI
International Water Management Institute

water witness

Water Stewardship Initiative

EWP
The Voice of Water

Marketplace Drivers: Global Water Roundtable

ALLIANCE FOR WATER STEWARDSHIP

ABOUT AWS | BOARD ORGANIZATIONS | WATER ROUNDTABLE | REGIONAL INITIATIVES | REGISTER INTEREST

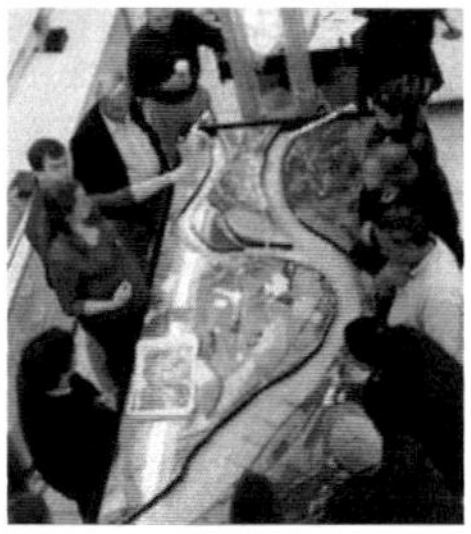

- Created to Develop Water Sustainability Standard by July 2013
- Basis for Water Certification Program
- Significant Stakeholder Input
  - Region
  - Sectors

BETTER PRACTICES BETTER PLANET 2020
Continuing AF&PA's Commitment to Sustainability

14

Marketplace Drivers: Disclosure and Risk Reduction Initiatives

- Similar to Carbon
- Water Use/Management Can Create Enterprise-Wide Risks (e.g., Coke in India)
- Taking the Time to Gather and Report Data will Result in Better Understanding of the Risks
- Important Information for Investors

BETTER PRACTICES BETTER PLANET 2020
Continuing AF&PA's Commitment to Sustainability

15

Marketplace Drivers: Disclosure

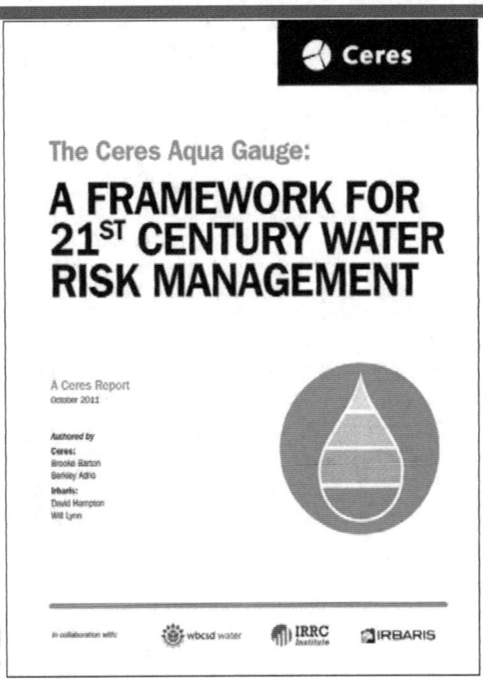
Ceres

The Ceres Aqua Gauge:

A FRAMEWORK FOR 21ST CENTURY WATER RISK MANAGEMENT

A Ceres Report
October 2011

Authored by
Ceres:
Brooke Barton
Berkley Adrio
Irbaris:
David Hampton
Will Lynn

In collaboration with: wbcsd water, IRRC Institute, IRBARIS

BETTER PRACTICES BETTER PLANET 2020
Continuing AF&PA's Commitment to Sustainability

16

Marketplace Drivers: Customer Water Goals

Customer	Water Goals
Colgate, Inc	-Reduce water use per ton of production by 40% by 2010 from 2002 levels -Reduce wastewater contaminants by 30% by 2010 from 2002 levels
General Mills	-Reduce usage 5% by 2011 from 2006 levels
PepsiCo.	-Reduce water use 20% by 2015 from 2006 baseline
Proctor and Gamble	-Reduce water consumption 20% by 2012 from 2002 levels
Starbucks	-Reduce water usage 25% by 2015 from 2008 baseline (also: 15% by 2012)
Unilever	-Maintain water abstraction levels to 2008 levels or 78% reduction per ton and 65% absolute reduction from 1995 baseline
The Coca-Cola Company	-20% water efficiency improvement by 2012 from 2004 baseline

BETTER PRACTICES 2020
BETTER PLANET
Continuing AF&PA's Commitment to Sustainability

17

Driver: Drought

- Drought in Southeast US in 2008 ->
- "The forest products sector is particularly susceptible to scarcity-induced disruptions or increased costs for water and/or energy supplies because it is very water- and energy intensive, especially in pulp and paper manufacturing." - Water Scarcity and Climate Change Report, Pacific Institute
- In 2007, at least four large pulp mills on the Alabama and Black Warrior rivers had to suspend wastewater discharges for days at a time to avoid violating their discharge permits. - NOAA

BETTER PRACTICES 2020
BETTER PLANET
Continuing AF&PA's Commitment to Sustainability

18

Industry Response

- Data Collection and Reporting Progress
- Commitments
- Engaging in the Development of the Future of Water Sustainability

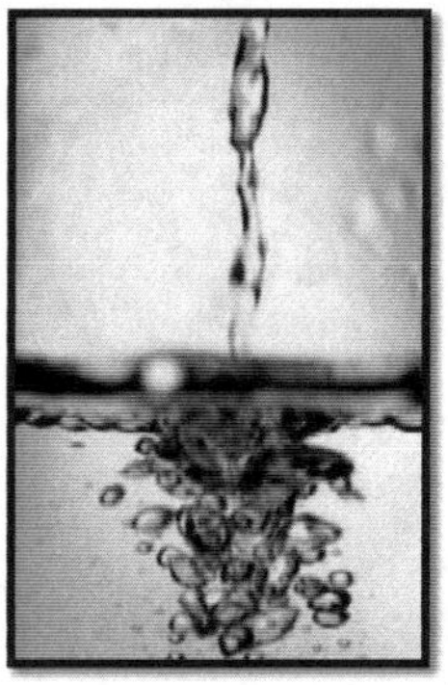

BETTER PRACTICES BETTER PLANET 2020
Continuing AF&PA's Commitment to Sustainability

19

Effluent Discharge Volume Decreased By Over 50% (1972-2008) and 20% Between 1995 and 2008

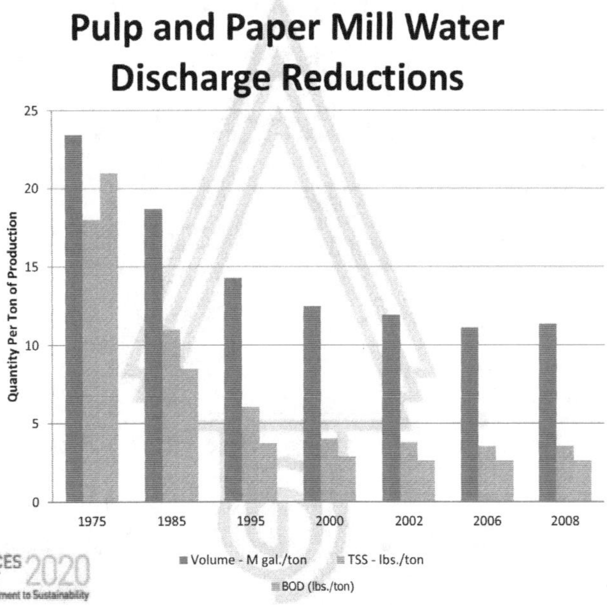

BETTER PRACTICES BETTER PLANET 2020
Continuing AF&PA's Commitment to Sustainability

Industry Response: Goals

Pillar	Goal (all Goals have a 2020 target and a 2005 baseline, except safety, which has a 2006 baseline)
Paper Recovery for Recycling	Exceed 70%.
Energy Efficiency	Exceed 10% increase. Measurement consistent with DOE method that rewards energy self sufficiency.
Greenhouse Gases	Exceed 15% reduction.
Forestry Related Goal	AF&PA members are committed to sustainable forestry both in the US and around the globe. We will increase the amount of fiber procured from certified forest lands or through certified fiber sourcing programs in the United States; and work with governments, industry and other stakeholders to promote policies around the globe to decrease illegal logging.
Safety-Recordable Case Rate	"Aspirational" goal of zero. Report and measure interim target of a 25% reduction in OSHA recordable incidents.
Water Use	Commit to collect and analyze data to address consumptive use to determine feasibility of later setting a numeric goal.

BETTER PRACTICES BETTER PLANET 2020
Continuing AF&PA's Commitment to Sustainability

21

Water Use

- Commit to Collect and Analyze Data to Address Consumptive Use to Determine the Feasibility of Later Setting a Numeric Goal

BETTER PRACTICES BETTER PLANET 2020
Continuing AF&PA's Commitment to Sustainability

22

AF&PA 2010 Sustainability Report

BETTER PRACTICES 202
BETTER PLANET
Continuing AF&PA's Commitment to Sustainab

23

Making a Public Commitment

BETTER PRACTICES 2020
BETTER PLANET
Continuing AF&PA's Commitment to Sustainability

24

INTERNATIONAL COUNCIL OF FOREST & PAPER ASSOCIATIONS

ICFPA AND WATER SUSTAINABILITY

World Water Week Stockholm

Jerry Schwartz
AF&PA
August 25, 2011

Water Sustainability – Global Engagement

- Goal: Outcomes the Industry Can Support
- AWS Global Roundtable
  - ISDC
  - North American Regional Council
- Water Footprint
  - Review
  - Pilots
- US Great Lakes Pilot Study of Various Initiatives

BETTER PRACTICES BETTER PLANET 2020
Continuing AF&PA's Commitment to Sustainability

Great Lakes Industry Study

- High Volume Withdrawal/Low Consumptive Use Industries in Water Rich Region
- Water Use Creates Economic Value
- Sponsors: Great Lakes Protection Fund, Council of Great Lakes Industries, NCASI, EPRI
- Key Input into Developing Initiatives

BETTER PRACTICES BETTER PLANET 2020
Continuing AF&PA's Commitment to Sustainability

Conclusions

- Forest Products Industry Water Sustainability is Moving Beyond Water Quality to:
    - Water Quantity
    - Impacts of Water Use
- The Industry is Committed to Better Understanding its Water Impacts & is Participating in Global Initiatives to Develop Water Sustainability Tools and Programs.
- Any Water Sustainability Tool Must be Able to Reflect the Site-Specific Nature of Water Resources and the Applicable Regulatory Environment.

BETTER PRACTICES BETTER PLANET 2020
Continuing AF&PA's Commitment to Sustainability

ALLIANCE FOR WATER STEWARDSHIP

From Local to Global Water Stewardship: Details on the Alliance for Water Stewardship

Nicole Tanner
WWF-US / Alliance for Water Stewardship

Water Sustainability in the Pulp & Paper Industry Conference
November 10th, 2011

CARBON DISCLOSURE PROJECT · The CEO Water Mandate · EWP · IWMI International Water Management Institute · The Nature Conservancy · Pacific Institute · Water Stewardship Initiative · water witness · Water Environment Federation · WWF

Agenda

ALLIANCE FOR WATER STEWARDSHIP

- **External Pressures and Drivers-Global**
- **External Pressures and Drivers-Local**
- **Water Stewardship and what it means for you**

The World of Today: Water Scarcity

ALLIANCE FOR WATER STEWARDSHIP

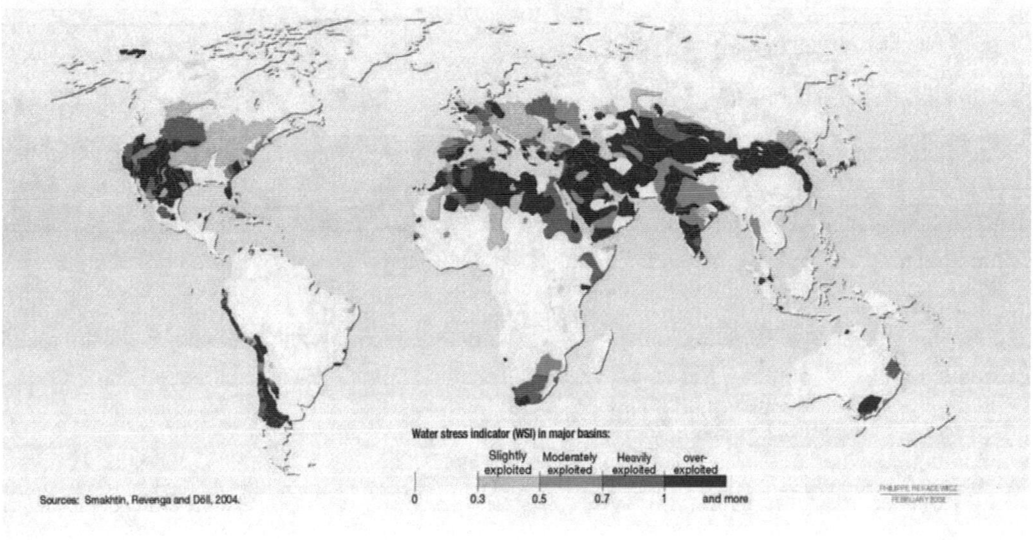

The World of Tomorrow: Climate, Consumption & Population Pressures

ALLIANCE FOR WATER STEWARDSHIP

- Changing patterns & availability:
  - Precipitation
  - Meltwater
  - Infiltration/overland flow
  - Baseflow
  - Groundwater
- Changing demand
  - Land use demands
  - Blue water demands (energy)
  - Green water demands (food)
  - Grey water demands (pollution)

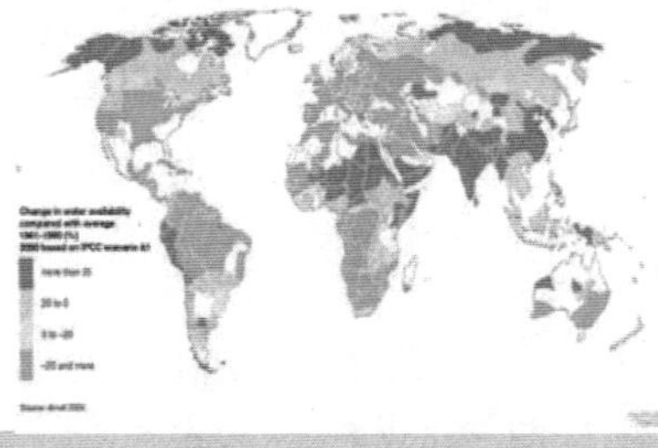

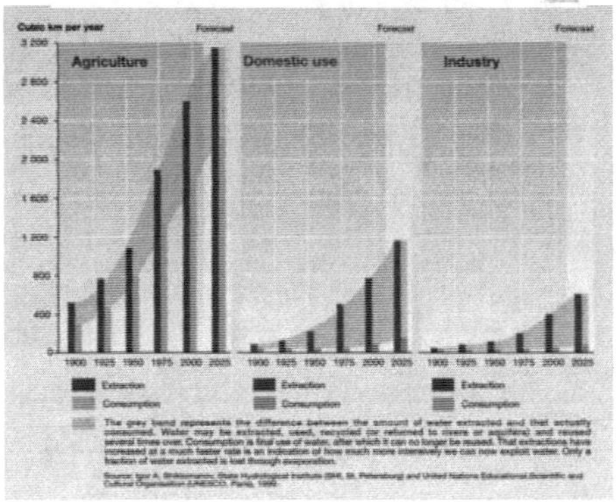

Increased Public Concern

ALLIANCE FOR WATER STEWARDSHIP

People around the world identify water issues as the most serious sustainability challenges facing the planet

GlobeScan-Circle of Blue survey (2009)

- 32,000 people 15 countries
- More than 90% perceived “water pollution” and “freshwater shortage” as serious problems, (70 percent “very serious”)
- Water concerns outpaced concerns about other well-recognized sustainability challenges such as global climate change, natural resource depletion, and biodiversity loss

Attitudes toward Water Issues

circle of blue

Average of 7 Countries, 2009

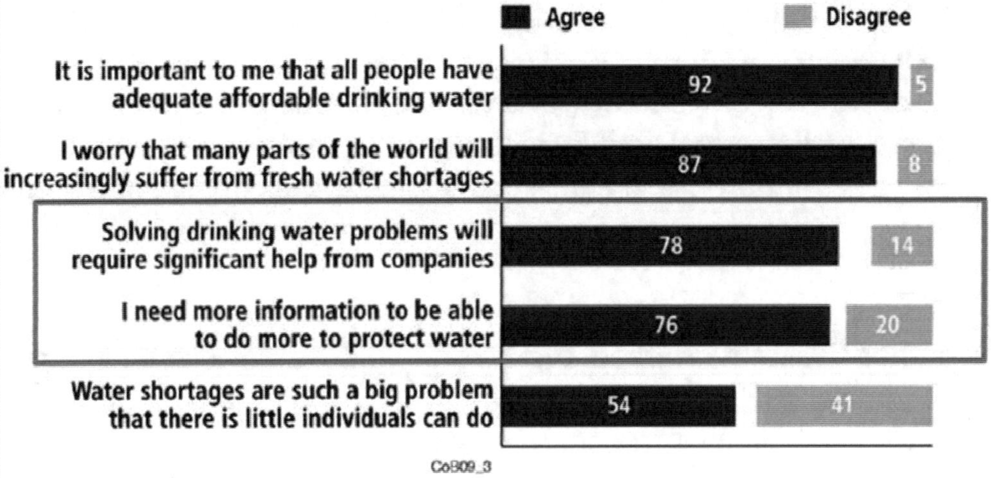

Emerging Shareholder Expectations

ALLIANCE FOR WATER STEWARDSHIP

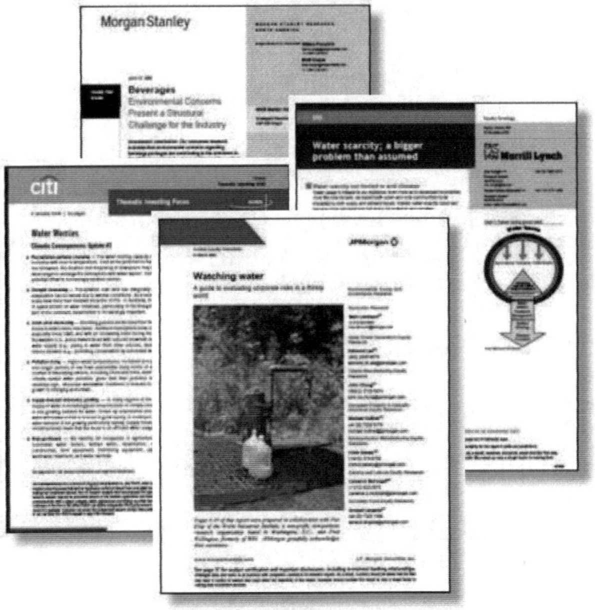

"Corporate disclosure of water-related risks is seriously inadequate and is typically included in environmental statements prepared for public relations purposes rather than in the regulatory filings on which most investors rely."

JP Morgan Global Equity Research March 2008

U.S. Securities Exchange Commission Guidance on Climate Related Disclosures

ALLIANCE FOR WATER STEWARDSHIP

D. Physical Impacts of Climate Change
Significant physical effects of climate change, such as effects on the severity of weather (for example, floods or hurricanes), sea levels, the arability of farmland, and water availability and quality, have the potential to affect a registrant's operations and results.

...Registrants whose businesses may be vulnerable to severe weather or climate related events should consider disclosing material risks of, or consequences from, such events in their publicly filed disclosure documents

Nature of Risk

ALLIANCE FOR WATER STEWARDSHIP

Depends on:

- The industry sector / nature of water use
- The maturity and reach of internal practice
- Hydrologic context (i.e. water scarce)
- Political context (i.e. sufficient capacity)
- Economic context (i.e. developing country)
- Social context (i.e. access to water/sanitation)
- Environmental context (i.e. healthy ecosystems)

The actions/conditions that create risk are often those that companies have least influence over

Water-Related Risk In the Value Chain

Point of impact: / Type of risk:	Supply chain	Production process	Product use
Physical	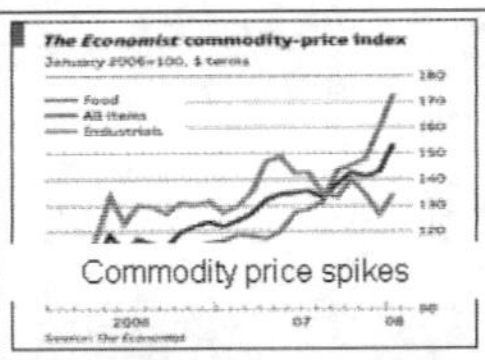 Commodity price spikes	 Disruption in water supply	 Scarcity limiting sales
Regulatory (+ litigation)	 Water quality standards constraining power generation	 Court settlement to scale back operations	 Insecure water rights
Reputation	 Multinationals' suppliers singled out for violations	Competition with social uses 	Profligate water use 

Source: *Treating Water*, April 2, 2009, Robeco in collaboration with WRI

What does it mean for businesses?

- Water procurement costs (supply-demand)
- Physical availability risks (no water)
- Political availability risks (water, no access)
- Environmental damage costs (social & environmental remediation)
- Reputational risks (brand damage)
- Preferred status / procurement (supply chain)
- Financial risks (water risk as a material concern of institutional investors, lending rates)

Watershed Level

- Water Stewardship
  - Cross industry
  - "Watershed-focused"
  - Cumulative impact

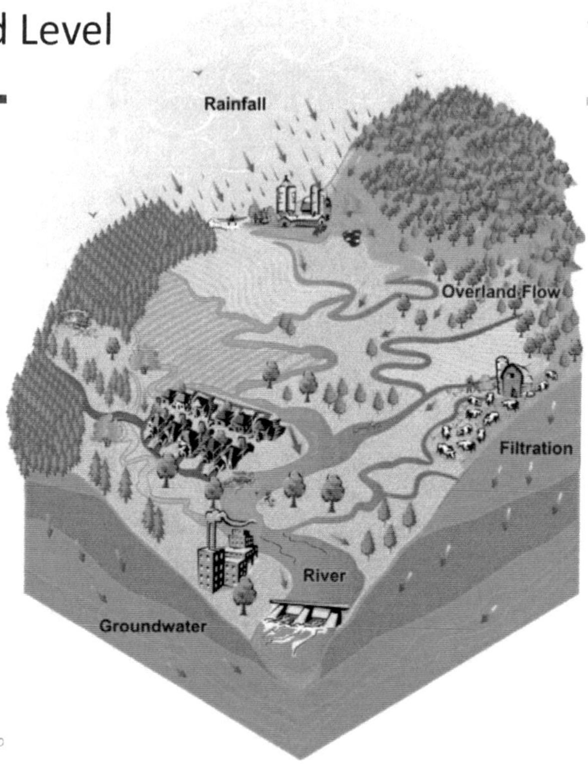

Shared Risk

ALLIANCE FOR WATER STEWARDSHIP

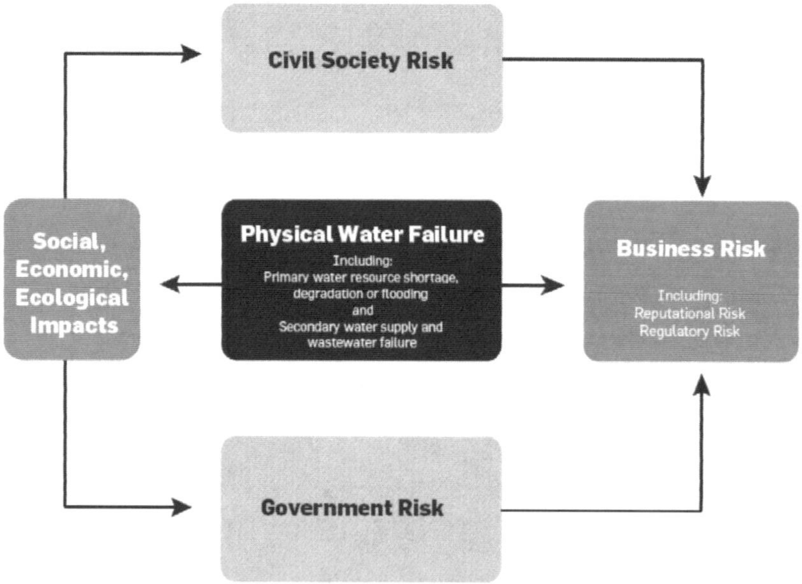

	Physical risks	Regulatory risks	Reputation risks
Caused by your own operations			
Caused by other stakeholders in water basin			

TBD-Commodity info to be Forestry related

Basin

- **Water Supply**
  - Groundwater status and interaction
  - Surface water status
  - Precipitation
- **Water Demand**
  - Ag Percentage of Total Demand
  - Water Stress/Scarcity
- **Water Quality**
  - Surface Water status
  - Groundwater status
- **Ecosystem Health**
  - Freshwater Ecoregion Status
  - Freshwater Biodiversity
  - Endangered/Threatened/Endemic Species
  - Locally Significant Features
  - Current Threats

Commodity

- **Water Demand**
  - Commodity Percentage of Total Demand
  - Commodity Water Requirements
    - Irrigation
    - Animal maintenance
- **Water Quality**
  - Agrochemical application rate and toxicity
  - Fertilizer application rate
  - Erosion/sedimentation impact
  - Manure
- **Ecosystem Health**
  - Commodity impact on ecosystem health

16

TBD

- **Future Impacts**
  - Water Supply: *Climate Change Impact*
  - Water Stress: *Trends in use, population, etc.*
  - Water Quality: *Trends in regulation, commodity relationship to quality, etc.*
  - Ecological Impact: *Trends in threats*

17

Risk definitions

Physical risk	Relates to water quantity (scarcity and flooding), water quality that is unfit for use (pollution), the impact of both quality and quantity on the surrounding ecosystem.
Regulatory risk	Relates to the imposition of restrictions on water use by government. This may include the pricing of water supply and waste discharge, licenses to operate, water rights, quality standards etc.
Reputation risk	Relates to the impact on a company's brand and that can influence customer purchasing decisions. Manifests through tensions and conflict around access to water or the degradation of local water resources. In a highly globalised information economy, public perceptions can emerge rapidly around business decisions that are seen to impact on aquatic ecosystems or local communities' access to clean water.

18

Quantifying Risk

ALLIANCE FOR WATER STEWARDSHIP

- Physical risk
- Regulatory risk
- Reputational risk
  f(Physical & Regulatory)
- Financial risk
  f(Physical, Regulatory & Reputational)

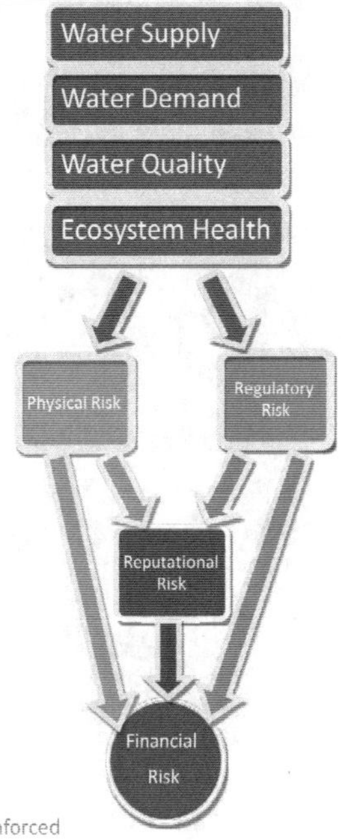

19

How have Businesses Responded?

ALLIANCE FOR WATER STEWARDSHIP

Voluntary action:

- Proliferation of water risk tools & approaches

AQUEDUCT Measuring and Mapping Water Risk
Global Reporting Initiative
GEMI Global Environmental Management Initiative
Global Water tool
Water Footprint NETWORK
CARBON DISCLOSURE PROJECT

- Standards

ISO International Organization for Standardization
ISEAL ALLIANCE

- Partnerships

WORLD RESOURCES INSTITUTE
The Nature Conservancy
CONSERVATION INTERNATIONAL
WWF
The CEO Water Mandate

- AWS

ALLIANCE FOR WATER STEWARDSHIP

Thank you

ALLIANCE FOR WATER STEWARDSHIP

The Alliance for Water Stewardship

ALLIANCE FOR WATER STEWARDSHIP

The Nature Conservancy — Protecting nature. Preserving life.

Pacific Institute

WWF

water witness

Water Stewardship Australia

Water Environment Federation — the water quality people

IWMI International Water Management Institute

The CEO Water Mandate

EWP The Voice of Water

CARBON DISCLOSURE PROJECT

My story

- Born and raised in the Finger Lakes region of upstate New York, outside of Syracuse but now live in Washington, DC.

- Education: BS in Political Science and International Relations (Elmira College, NY), MA in International Development – concentrated in sustainable development and resource based conflict (DU's Korbel School of International Studies)

- Work: WWF-US/AWS for 3.5 years on Market Transformation and Freshwater issues

Water Sustainability in the Pulp and Paper Industry

The CGLI Water Footprinting Project

Barry Malmberg

NCASI

Drivers for Water Sustainability and Stewardship Programs

- water scarcity
  - UN World Water Development Report: *Water in a Changing World (2009)*
- water use and water availability
  - concern over water transfers, both 'real' and 'virtual'
- corporate environmental reporting
- an "opportunity for certification"

Water Initiatives

sustainability / stewardship / footprint

- Aquawareness
- Alliance for Water Stewardship
- Blue Certification Program
- Carbon Disclosure Project for water
- Corporate Water Gauge
- European Water Partnership
- Federation House Commitment to Water Efficiency
- Global Environmental Management Initiative
- Global Reporting Initiative
- Global Water Tool (GWT)
- UNEP/SETAC Life Cycle Initiative
- Strategic Water Management Framework (Australia minerals)
- UN CEO Water Mandate
- Water Brief for Business SEE Initiative
- Water Footprint Network
- Water Neutral Foundation
- WaterSense Certification Scheme
- Water Stewardship Initiative

CGLI Water Footprinting Project

- Council of Great Lakes Industries' (CGLI) water footprinting project
  - Evaluate major water sustainability initiatives
  - Application of water sustainability initiatives to industries important to the Great Lakes – St. Lawrence River Basin
- Answer questions such as
  - Are the water sustainability initiatives applicable to "water rich" regions like the Great Lakes Basin?
  - Do the initiatives appropriately consider industry water use needs?
  - Will the initiatives affect future water use approvals and impact costs?

Review of General Approach

- Two alternative approaches to address study questions:
  - "Whole initiatives" approach
    - Examples: Water Footprint Network, Global Water Tool
  - "Common metrics" approach
    - Examples: Total volume consumed, consumption relative to total available supply

"Whole Initiatives" Approach

- Approach: Select and apply 2-3 initiatives to each facility
- Advantages:
  - Simpler to understand and apply
- Disadvantages:
  - Have to select initiatives up front
  - Will not originate from a Great Lakes perspective
  - Would not consider full range of initiatives
  - May miss relevant metrics

"Common Metrics" Approach

- Approach: Select common and relevant metrics and apply separately to each facility, then link key metrics back to initiatives
- Advantages:
  - Addresses the primary elements in many initiatives
  - Focuses on specific relevant measures, allowing for a more focused Great Lakes evaluation
  - Avoids "taking sides" on specific initiatives
- Disadvantages:
  - Requires multi-step approach
  - May be harder to understand

Withdrawal and Source Metrics

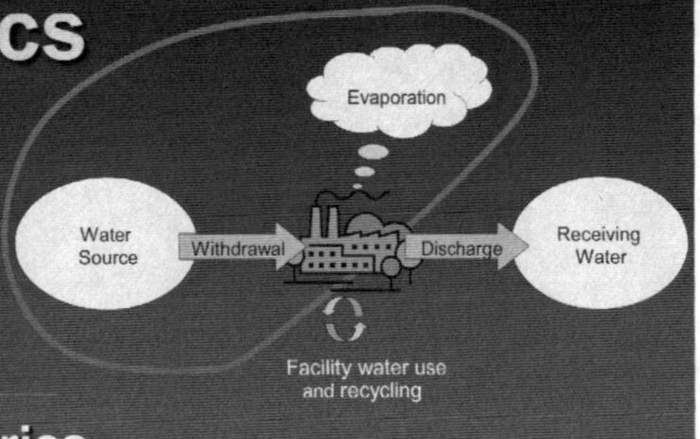

- Withdrawal amounts
  - Volume withdrawn by source
  - Volume consumed by source
  - Water transfers
  - Peak/average/seasonal use
- Source characterization metrics
  - Sources under stress
  - Amount of renewable water
- Withdrawal impact and available supply
  - Relative to total available supply
  - Effect on ecosystem services
  - Effect on human services

Discharge and Recycling Metrics

- Discharge amounts
  - Volume discharged to receiving body
- Discharge quality
  - Regulated pollutant load
  - Non-regulated pollutant load
- Discharge impact
  - Eutrophication potential
- Effect on ecosystem
  - Effect on downstream human uses
- Recycling/Reuse
  - Internal recycling and reuse
  - External recycling and reuse

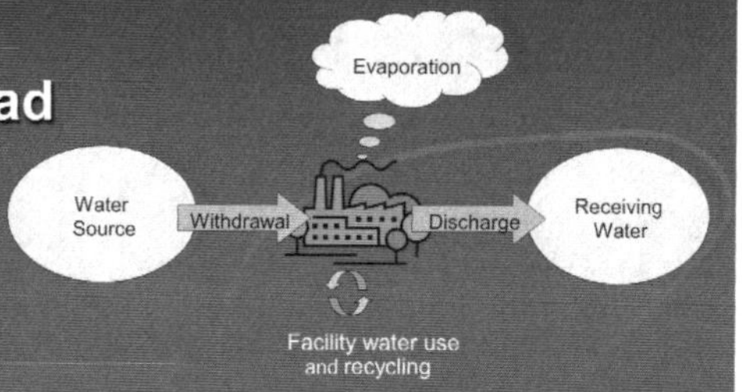

Governance and Benefit Metrics

- Equitable and Transparent Governance
  - Water consumption per unit product
  - Water resource management strategy (use and disclosure)
  - Permits and other consents – withdrawals
  - Permits and other consents - discharges
- Benefits
  - Economic and social benefits

Metrics Application

- Determining the scale for metric application
- Compile needed data and information
  - Survey questions
  - Measurements of water quality and quantity
  - Maps and diagrams
  - Supporting facility documentation (e.g., BMPs)
  - Outside assessments, calculations and professional opinions
- Develop a facility water balance

NewPage Escanaba Pulp and Paper Mill

- General Description
  - Located off shore of Lake Michigan
  - Located just north of the town of Escanaba, MI – population ~13,000
  - About 1,100 site workers
  - Products produced – Coated freesheet from kraft pulp (77%), coated groundwood from refiner mechanical pulp (20%), and market pulp (3%)
  - ~675 million $ sales/year

NewPage Escanaba Mill
Escanaba Mill

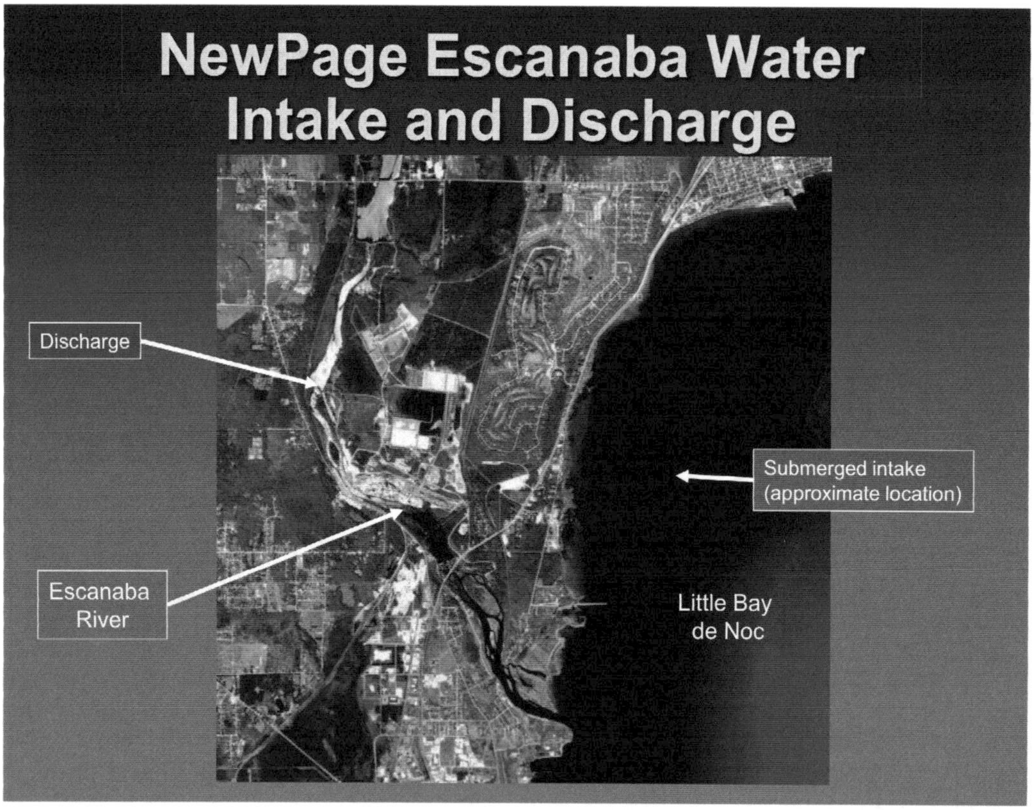
NewPage Escanaba Water Intake and Discharge
Discharge
Submerged intake (approximate location)
Escanaba River
Little Bay de Noc

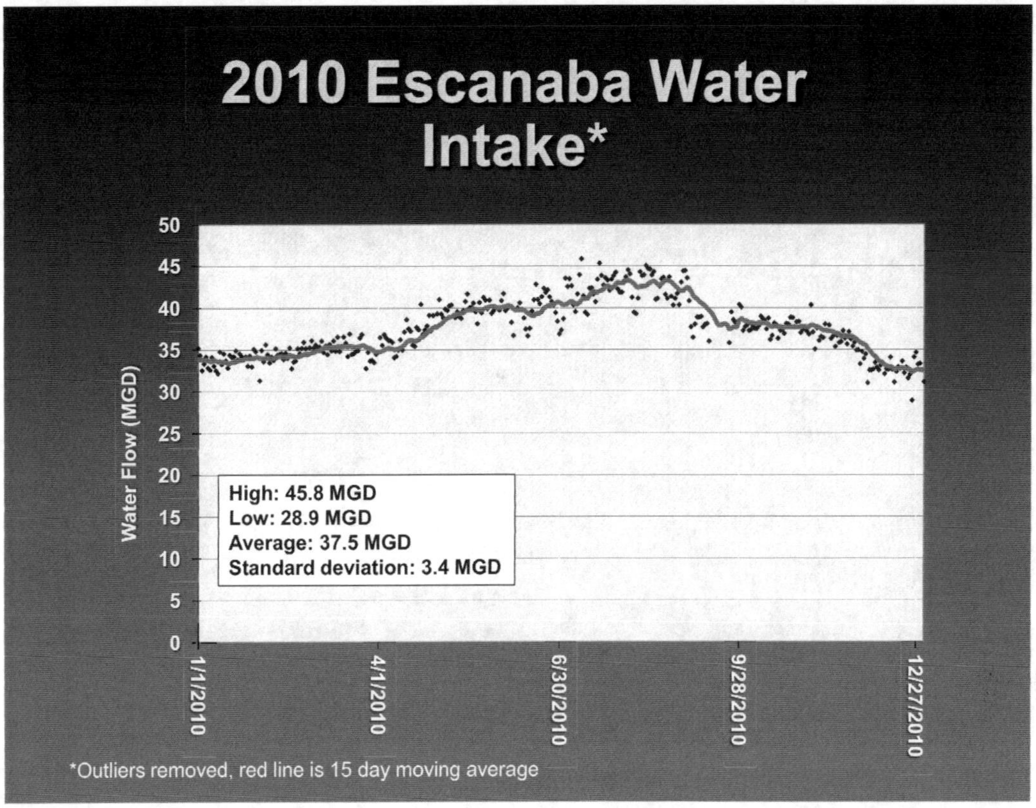
2010 Escanaba Water Intake*
Water Flow (MGD)
50
45
40
35
30
25
20
15
10
5
0
High: 45.8 MGD
Low: 28.9 MGD
Average: 37.5 MGD
Standard deviation: 3.4 MGD
1/1/2010
4/1/2010
6/30/2010
9/28/2010
12/27/2010
*Outliers removed, red line is 15 day moving average

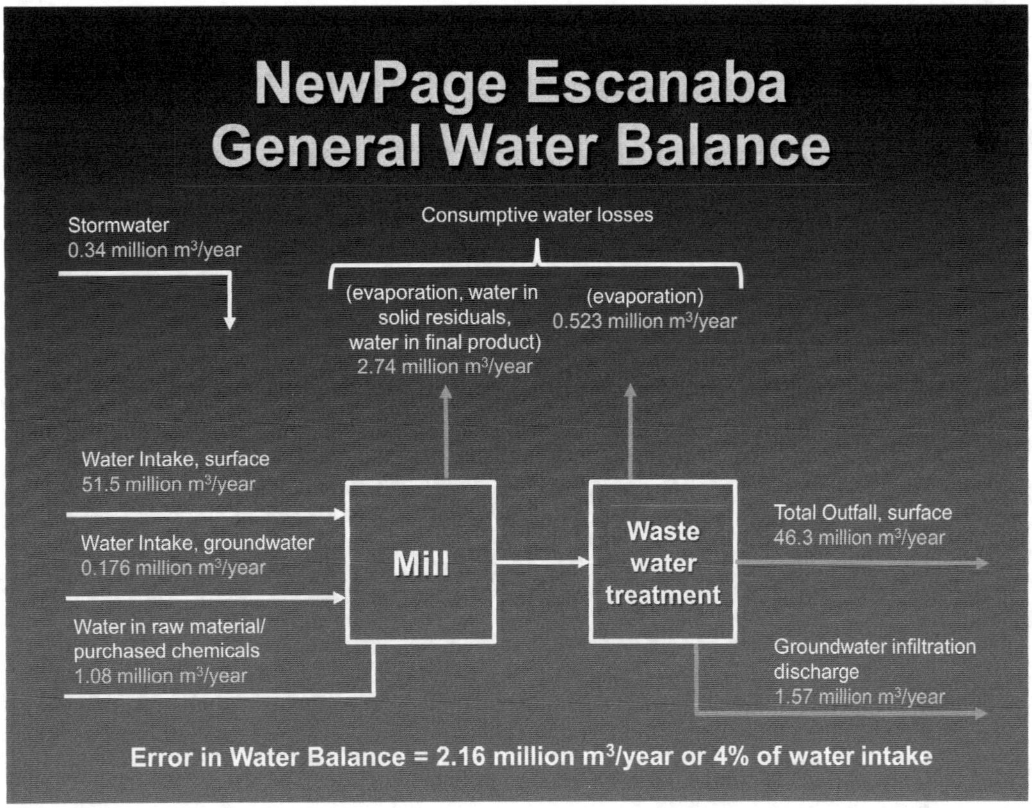
NewPage Escanaba General Water Balance
Stormwater
0.34 million m³/year
Consumptive water losses
(evaporation, water in solid residuals, water in final product)
2.74 million m³/year
(evaporation)
0.523 million m³/year
Water Intake, surface
51.5 million m³/year
Water Intake, groundwater
0.176 million m³/year
Water in raw material/ purchased chemicals
1.08 million m³/year
Mill
Waste water treatment
Total Outfall, surface
46.3 million m³/year
Groundwater infiltration discharge
1.57 million m³/year
Error in Water Balance = 2.16 million m³/year or 4% of water intake

Internal Recycling and Reuse

- Pulp and paper mills use extensive recycle of internal process streams for energy, water, and chemical conservation
- Internal water recycle at Escanaba is estimated to be 620 million m^3/year, approximately 12 times the annual water usage

Water Consumption

- Global Water Tool: (withdrawal – discharge)
  - 5.24 million m^3/year
  - No consideration of other water import and export vectors that affect the water balance
- Summation of consumptive water losses
  - 3.26 million m^3/year
  - evaporation + water in solid residuals + water leaving with product
- Difference in water consumption results vary by 60% depending upon the method of calculation

Water Footprint Network

Water Footprint

- Green Water: volume of evapotranspiration from agricultural or forest crops
- Blue Water: volume evaporated or otherwise lost from the manufacturing process
- Grey Water: volume required to dilute pollutants to achieve local standards for water quality

→ Water Footprint

Grey Water Footprint

discussion of pollutant loads

- **Escanaba has BOD_5, TSS, and TP discharge limits and an ammonia reporting requirement as part of their NPDES permit**
- **Escanaba conducts toxicity testing annually as part of the NPDES permit**
- **Escanaba performed an assimilative capacity study for BOD_5 and TP discharges**
- **Escanaba's effluent temperature is greater than the receiving water – monitoring requirements are in place as part of their NPDES permit**

Grey Water Footprint

Receiving Water Data

- **USGS Monitoring Stations**
  - **Cornell, MI 7 miles upstream**
    - flow and temperature data, last speciated N and P measurements were in 1994
  - **Escanaba River at Wells, MI, directly downstream of the mill**
    - Flow and some speciated N and P measurements

Grey Water Footprint

- **Water Footprint Network methodology: select the pollutant that generates the largest grey water footprint**
- **General lack of receiving water data for relevant pollutants, especially during relevant time periods. Preliminary grey water footprint based upon thermal discharge with temperature change guidance from WFN Manual, 111 million m^3/year or 215% of Escanaba's annual discharge**
- **Grey water footprint for total phosphorus – 59 million m^3/year**

Green Water Footprint

- *Methodology: The Green and Blue Water Footprint of Paper Products: Methodological Considerations and Quantification.* Research Report Series No. 46
- Escanaba fiber sources HW: 98% from the Upper Peninsula (UP), 2-3 from WI. SW: 80-90% from UP and 10-15% from Lower Michigan.
- Fiber sources: 40% from small private land owners, 35% from large industrial land owners, 20% from state and 5-10% from federal lands

Preliminary NewPage Escanaba Water Footprint Network Water Footprint

WFN Methodology

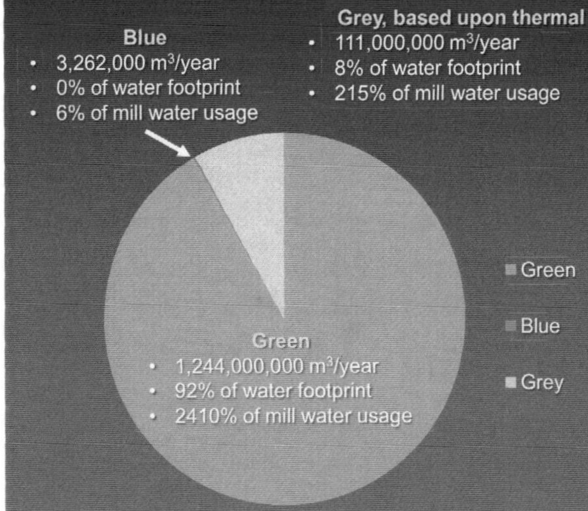

Valid Alternative Water Footprint

Grey water footprint = 0
Escanaba discharge meets all NPDES discharge limits

Net green water footprint ≈ 0
Land area for Escanaba wood fiber would be timber land under natural usage scenario

Blue
- $3,262,000\ m^3$/year
- 100% of water footprint
- 6% of mill water usage

Withdrawal/Discharge Impacts

- Escanaba withdrawal is 0.0006% of Lake Michigan volume
- No reported effects upon ecosystem services
- No reported effects upon human services

Data Needs

- Facility data requirements are significant. Facility cooperation is critical for any credible water footprinting exercise
- Receiving water data needs are considerable and often don’t coincide in type and timing with relevant mill-specific data

CGLI Water Footprinting Project

- Last Expert Panel Meeting was Nov. 9th.
- Final report will be complete in the first quarter of 2012.
  - Deficiencies in current water sustainability/stewardship initiatives
  - Recommendations for improvement to these programs
  - Recommendations for approaching water sustainability within the Great Lakes region

Challenges in Reducing Water Use

Presentation to

AIChE Water Sustainability in Pulp and Paper Workshop

November 10, 2011

Ron Brown
ron_brown@agenda2020.org

Agenda 2020 Technology Alliance

- Mission . . .
  - Identify R&D priorities that can help transform the industry – breakthrough R&D, not incremental
  - Promote collaborative R&D programs that address the industry's technology priorities
  - Seek funding to support R&D programs
- Partnership – industry/government/universities
- Incorporated as independent non-profit in 2011
  - Started 1994 – Became self-supporting in 2005
  - Located with AF&PA in DC

Agenda 2020 Members

Company Members

- ArborGen
- Fibria
- Georgia-Pacific
- Imerys
- MeadWestvaco
- Metso
- Nalco
- NewPage
- Sappi
- ThermoChem Recovery Intl
- UPM-Kymmene
- Verso Paper

Affiliates and Partners

- US Dept of Energy
- US Forest Service
- Forest Products Lab
- PPERA Universities
- IPST at Georgia Tech
- NCASI
- Bioenergy Deployment Consortium
- CleanTech Partners
- American Forest & Paper Assn
- TAPPI
- NPT2
- Herty Adv Matls Devt Ctr

2006 Technology Roadmap

- Six focus areas
  - Forest productivity
  - Forest biorefinery
  - Breakthrough manufacturing
  - Wood products
  - Environmental
  - Fiber recovery

- Water was not an area of focus
- Changes since 2006 indicated need for new roadmap
  - Sustainability, climate change, GHG emissions
  - Concerns about wood supply, energy, input costs, and water

Drafting a New Roadmap: Two Workshops

- Workshop 1 – What areas need breakthrough R&D?
  - 45 participants
  - Speakers addressed social issues and business challenges
- Workshop 2 – What technologies need to be developed?
  - 100 participants
  - In partnership with
    - IPST at Georgia Tech
    - DOE Industrial Technologies Program
    - American Forest & Paper Association

2010 Technology Roadmap

Six strategic areas

- Energy and GHG emissions
- Water
- Biomass supply
- New value streams from woody biomass
- New product features and novel materials
- Recovery and recycling of wastes

Much overlap with 2006 Roadmap

New areas – water, products

Download the Roadmap at www.agenda2020.org

2010 Technology Roadmap – Water

R&D Objectives

- Reduce fresh water used in pulping and papermaking by 50%
- Develop technologies to treat and reuse effluent in plants
- Develop closed-loop water systems

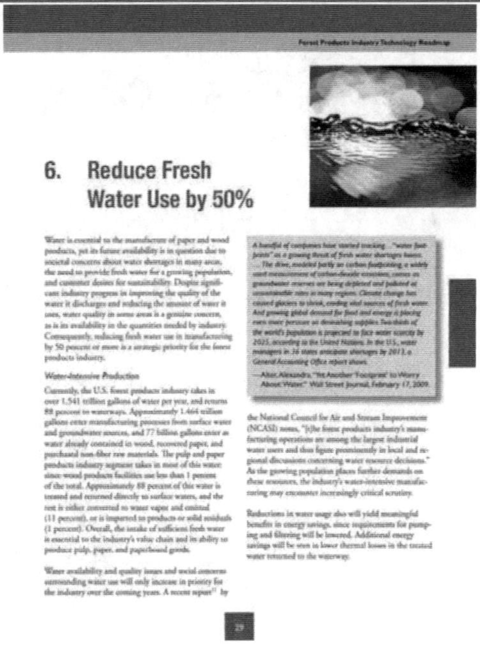
Forest Products Industry Technology Roadmap

6. Reduce Fresh Water Use by 50%

AGENDA 2020 TECHNOLOGY ALLIANCE *Transforming the forest products industry through innovation* 7

Roadmap – Water: Specific R&D Needs

- Separate dilute contaminants – organic, inorganic, and ions – from waste streams to enable reuse in place of fresh water
- Reduce fresh water used in pulp washing
- Remove non-process elements from chips before pulping
- Use biological treatment upstream in plant before waste treatment
- Improve process modeling tools – chemical equilibrium, ion tracking, life-cycle assessment
- Use heat recovery systems to better recover waste heat from process water streams
- Understand better how metals become fixed in wood
- Address scaling issues
- Learn from past experiences with closed water systems

AGENDA 2020 TECHNOLOGY ALLIANCE *Transforming the forest products industry through innovation* 8

2010 Roadmap Workshop: Water Group

- Companies
  - Ashland Hercules
  - BE&K
  - Eka Chemicals
  - International Paper
  - MeadWestvaco
- Government
  - Idaho National Lab
  - Forest Products Lab
- Universities
  - Georgia Tech
  - Miami University
  - North Carolina State University

2010 Roadmap: Selecting the Top R&D Priorities

- Six strategic areas, 27 technology objectives in Roadmap
- Progress on all objectives is desired for the good of the industry
- Which objectives need new technologies most urgently at the present time?
- Voting by Agenda 2020 members, partners, and other stakeholders
- 7 priorities for breakthrough R&D selected

Seven R&D Priorities in Four Platforms

- Sustainable Manufacturing
  - Reduce energy intensity of manufacturing by 25%
  - **Reduce fresh water intake in mills & plants by 50%**
- Value from Biomass
  - Develop new value-added products from woody biomass
  - Develop thermochemical biorefinery processes that are cost-effective for a typical pulp mill
- Sustainable Forest Productivity
  - Develop trees with high growth rates for specific end uses
- Novel Materials and Products
  - Improve strength-to-weight ratio of paper and paperboard by 20-50%
  - Develop new wood-derived nanomaterials and advanced composite structures

AGENDA 2020 TECHNOLOGY ALLIANCE *Transforming the forest products industry through innovation* 11

What Next for the Roadmap Priorities?

- Encourage research programs to address priorities
  - Collaborative, pre-competitive projects
    - Agenda 2020 as lead applicant with partners on some
  - Agenda 2020 task groups active
    - Virtual "circles of excellence" in technology area – e.g., separations – to identify best approaches and skills
- Seek government funding
  - Respond to funding opportunity announcements
    - Example – DOE Innovative Manufacturing Initiative
  - Influence future funding programs
    - Meetings to inform agencies and Congress of needs and social benefits
    - Participation in workshops and events

AGENDA 2020 TECHNOLOGY ALLIANCE *Transforming the forest products industry through innovation* 12

Reducing Water Use in Pulp and Paper

- Agenda 2020 goal is 50% reduction per ton for the industry average
- Best available technologies for new mills achieve this goal – examples in China
- Overall, good progress in last 30 years in lowering the industry average in the US
- Challenge: Reducing water use dramatically in existing mills

Water Use in the US Is Declining Slowly

Pulp and Paper Mill Water Discharge Reductions

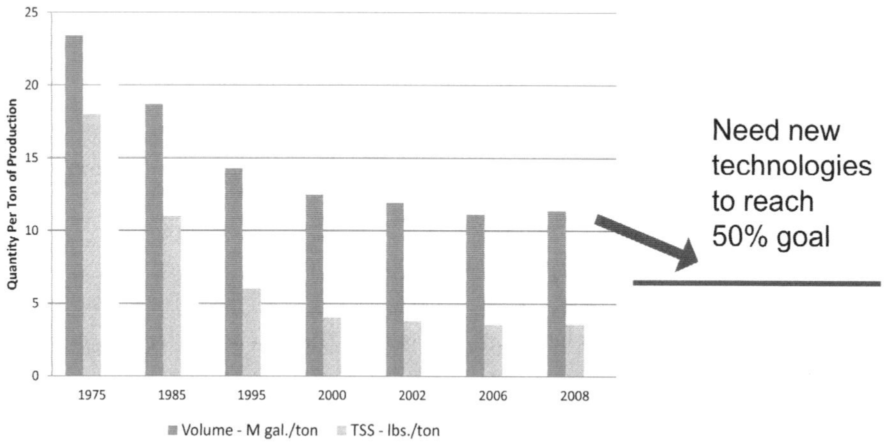

Source: AF&PA

Challenges to Achieving the 50% Goal in Existing Mills

- Economic and infrastructure
- Organizational, environmental, and social
- Technological

AGENDA 2020 TECHNOLOGY ALLIANCE *Transforming the forest products industry through innovation* 15

Challenges: Economic and Infrastructure

- Justifying new capital investment in existing mills
  - Capital funds are limited and often targeted at growth or to respond to regulations
- Realizing cost savings with new equipment
  - Successful demonstrations of new technologies to quantify cost benefits
  - Risks that changes will not work as planned
  - Unintended consequences
- Handling site-specific issues and situations
  - Sources of fresh water
  - Capacity of river to accept discharges

AGENDA 2020 TECHNOLOGY ALLIANCE *Transforming the forest products industry through innovation* 16

Challenges: Organizational, Environmental and Social

- Social and regulatory uncertainties
- Use of common language and terminology
  - Discharge – intake – consumption
- Organizational inertia
  - Little focus on improving basic mill systems
- Organizational reluctance to work with others and to use ideas from others
  - Other mills, other industries
- Leadership commitment
  - AF&PA Sustainability Goals – no specific water goal for 2020

AGENDA 2020 TECHNOLOGY ALLIANCE *Transforming the forest products industry through innovation* 17

Challenges: Technological

- Availability of adequate process data
- Risks of trying new methods
  - Potential adverse impact on products and processes
- Limited technical capacity in mills and on corporate staffs to support implementation of new technologies
- Low level of R&D funding for new technologies
- Lack of breakthrough innovations

AGENDA 2020 TECHNOLOGY ALLIANCE *Transforming the forest products industry through innovation* 18

R&D as Percent of Sales – 2008: Paper Industry is Low vs. Other Industries

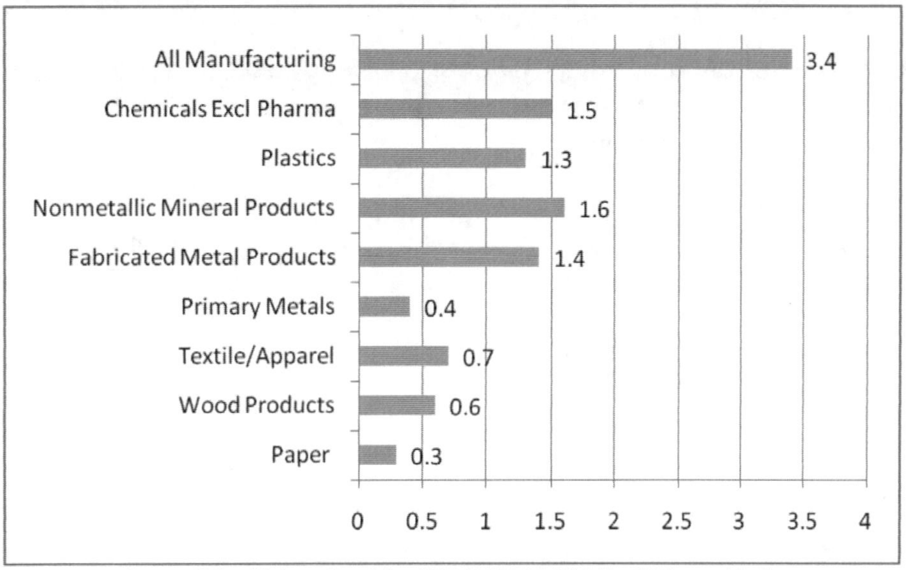

Source: National Science Foundation, NSF 10-322, InfoBrief, May 2010, except for Paper.
Paper data from company annual reports for 10 companies in U.S. for 2008.

Paper Industry R&D Has Declined Since 2006

In U.S. companies, R&D spending as % of sales declined one-third from 0.34% in 2006 to 0.24% in 2010.

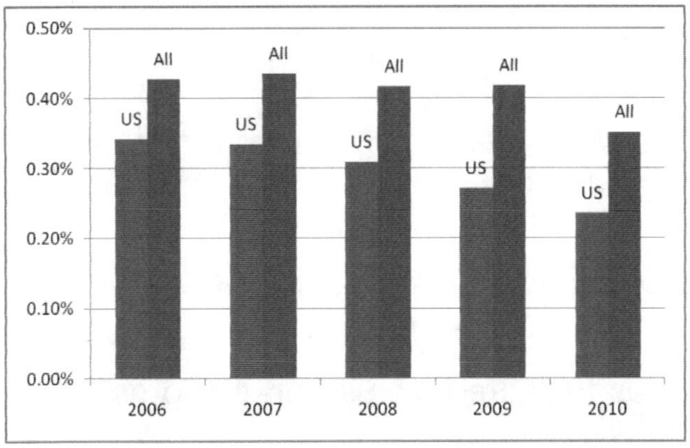

Data represent $95B sales in 2010 and were obtained from company annual reports:

All = 13 companies
US = 10 companies

In 2010, U.S. companies ranged from 0 to 0.7%.

US: Graphic Packaging, International Paper, MeadWestvaco, Pkg Corp America, Rock-Tenn, Sonoco, Smurfit-Stone, Temple-Inland, Wausau Paper, Weyerhaeuser. Outside US: Sappi, Stora Enso, UPM-Kymmene.

Promoting Innovation in the Paper Industry

- Low level of company R&D spending suggests that more collaborative programs are needed to address industry's technology priorities
- Collaborative programs financed in large part by government funds are a good format for pre-competitive R&D

Many Formats for Collaborative R&D

- Program-specific research consortium
  - Examples at NCSU, SUNY ESF, Auburn, U Maine, ...
- Industry-supported research institute
  - IPST at Georgia Tech
- Public/private partnership – industry/government/academia
  - FP Innovations – Canada
- Government initiatives
  - European Union 7th Framework Programme – Forest-Based Sector Technology Platform (FTP)
- Alliance of companies to promote collaborative programs
  - Agenda 2020 Technology Alliance

Path Forward for Innovation

Approaches to promote innovation in technologies:

- Increase R&D spending – not likely in the near term
- Conduct joint R&D programs
  - Use "open innovation" concepts and seek best partners
  - Good for company-driven initiatives for competitive advantage
- Work together to address industry-wide technology needs with pre-competitive R&D
  - Share the costs – developing new technologies is expensive
  - Benefit from government financial support
  - Involve best available ideas and people worldwide
  - Agenda 2020 provides an organized approach for collaborative development of new technologies

AGENDA 2020 TECHNOLOGY ALLIANCE *Transforming the forest products industry through innovation* 23

Summary: Challenges in Reducing Water Use

- The 2010 Technology Roadmap outlines R&D needs
  - Reducing water use by 50% is a top R&D priority selected by Agenda 2020 stakeholders
  - New technologies are needed particularly for existing mills
- Multiple challenges exist to reducing water use
  - Economic, environmental, social, organizational, technological
- Low R&D in paper industry hinders development of innovative approaches
- Collaborative R&D is a good way forward

AGENDA 2020 TECHNOLOGY ALLIANCE *Transforming the forest products industry through innovation* 24

Thank You for Listening

Ron Brown

ron_brown@afandpa.org

1-202-463-2742

NALCO

Essential Expertise for Water, Energy and Air

Innovation Needs for Water Re-Use in the Pulp and Paper Industry

Water Sustainability in the Pulp and Paper Industry Workshop
November 11, 2011

Water's Role in Pulp and Paper Production

- Water is a vital component in the production of pulp and paper.
- Water can impact the production efficiency and end product quality.
- A considerable amount of water is currently re-used or recycled in pulp and paper production.
- Technologies exist which have enabled mills to operate at "zero effluent". The majority of these involve large capital expense.
- Areas that need technology innovations to enable water re-use or reduction will be highlighted. The focus will be on monitoring and removal of contaminants to provide water that is "treated to need".

2

Example of Fresh Water Usage for Integrated and Non-integrated Mills

Integrated tissue mill example: pulping accounts for 40% of fresh water usage.

Non-integrated mill example: three paper machines account for 60% of fresh water usage.

Values based on withdrawals from fresh water tank

3

Common Water Contaminants in Paper Production and Related Concerns

Contaminants	Consequences
Chlorides, acidity (H^+ ions)	Corrosion
Sulfates	Scale
Polyphosphates	Loss in chemical additive performance
Calcium, aluminum, or magnesium	Scale, interference with chemical additives and cleaning products
Iron, manganese, lignin	Negative impact on brightness
Microbiological growth	Reduced machine efficiency
Organic materials (i.e. starch)	High conductivity; loss in chemical additive performance
Stickies, pitch colloids	Machine deposits, product defects

4

Water Quality Specifications for Paper and Pulp Products

Parameter	Maximum Recommended Concentration (mg / L)			
	Wood free Papers	Wood-containing Papers	Bleached Kraft Pulp	Unbleached Kraft Pulp
Turbidity as NTU	75	375	300	750
Total dissolved solids	200	500	300	500
Total hardness as $CaCO_3$	100	200	100	200
Calcium hardness as $CaCO_3$	50	NaV	NaV	NaV
Silica (soluble) as SiO_2	20	50	50	100
Iron as Fe	0.1	0.3	0.2	1.0
Manganese as Mn	0.05	0.1	0.1	0.5
Color, Pt/Co units	5	30	25	100
Residual chlorine as Cl_2	2.0	NaV	NaV	NaV
Chlorides as chlorine	NaV	75	200	200

5

Pulp and Papermaking Process

Examples of Technology Needs for Water Sustainability

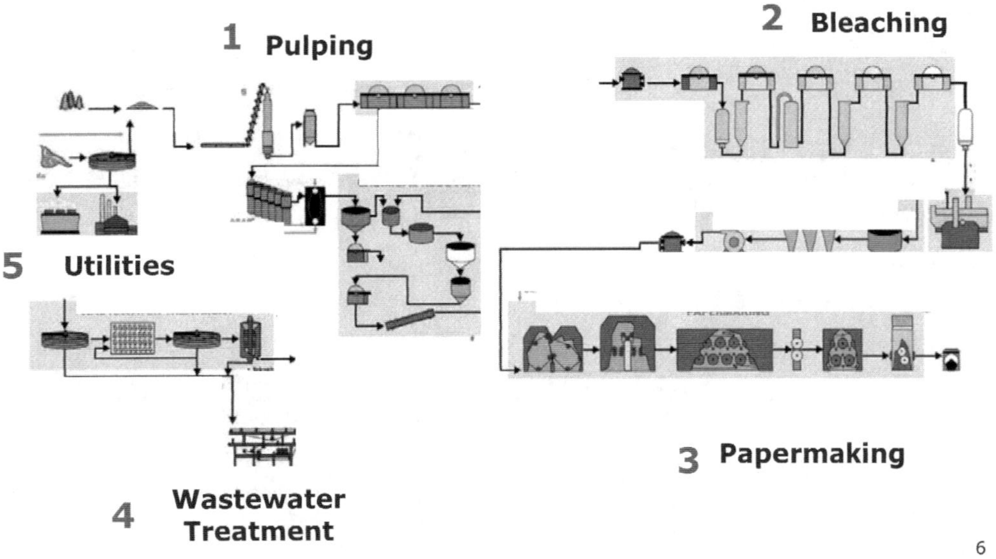

6

1 Technology Needs to Reduce Water Usage in Pulping

- Significant water is used to wash pulp after the various stages of pulping and bleaching.

- Technologies that enable washing to be completed at higher consistencies could reduce water consumption. Even an increase from 10% to 12% consistency would be beneficial.

- Examples of potential technical solutions:
  - Effective surfactant-based chemistries to facilitate removal of impurities
  - Improved mixing devices to accomplish washing at higher consistency

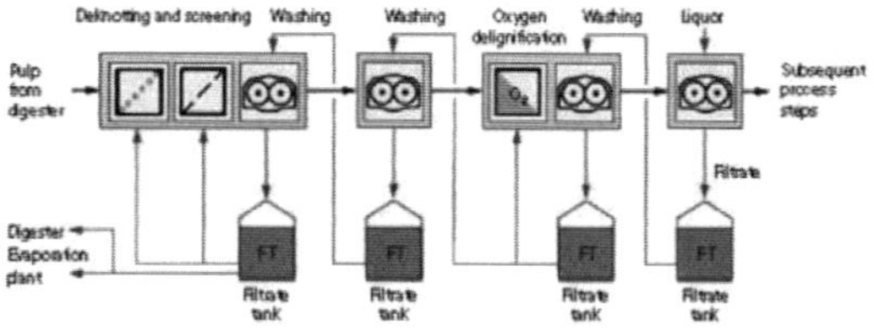

7

2 Technology Needs for Water Re-Use in Kraft Bleaching

- Bleaching is typically carried out at 10% pulp consistency and in many stages; the five common stages are $D_oE_{op}D_1ED_2$
  - Filtrate from the later bleaching stages flows counter current to serve as wash water for the previous step. Filtrate from the Do and Eop stages is commonly discharged to waste treatment
- First Chlorine Dioxide Stage Filtrate (D_o) Filtrate
  - Contaminants that cause scaling problems: barium, calcium, magnesium
  - Contaminants that degrade bleaching chemistry (peroxide): iron, manganese
- First Extraction Stage (E_{op}) Filtrate
  - Contaminants include colored materials (lignin), COD

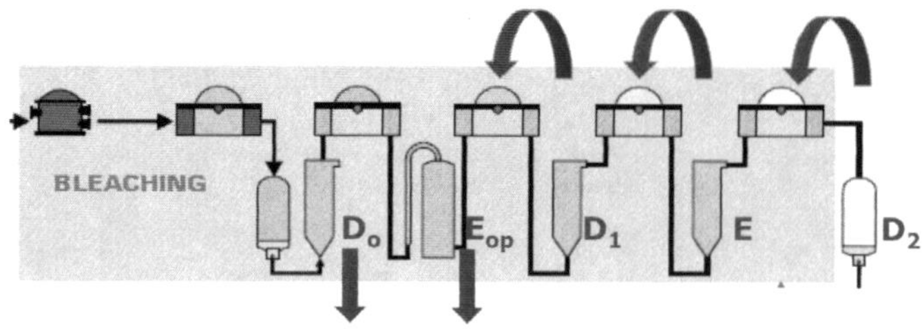

8

3 Papermaking: Re-use of Save-All Filtrate

- In general, water re-use is well managed in the paper machine.
- Potential for re-use is the discharge from the Save-All clear leg to the waste treatment plant.

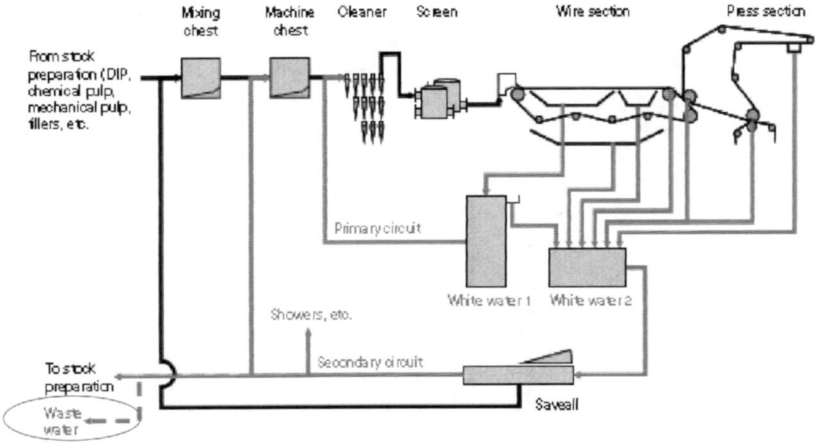

Figure 5 Schematic illustration of principal water and stock flows in a paper machine (Hutter, 2008).

9

3 Papermaking: Re-use of Save-All Filtrate

- Clear filtrate from the Save-All could be potentially treated and re-used for shower water or non-contact pump seal water.
  - Potential energy savings from cost associated with heating water
- Technology needs:
  - Methods to reliably monitor and control TSS levels and concentration of potential scaling ions.
  - "Treat to need"

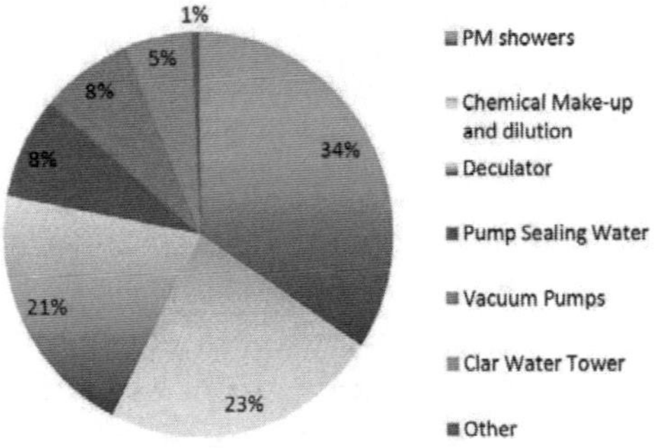

Distribution of Fresh Water Usage on a Paper Machine

10

4 Wastewater Discharge Re-Use in Papermaking Process

- Filtration techniques have been successfully implemented to purify water to acceptable levels for re-use in the process. These include UF, NF, and RO

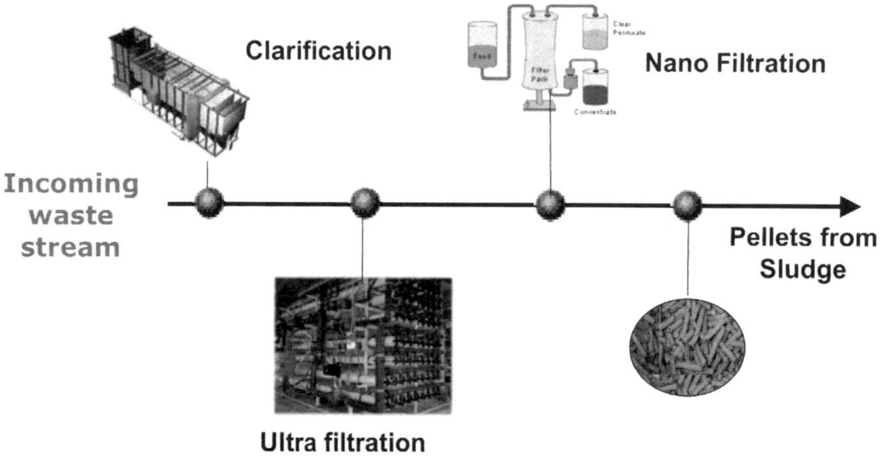

11

4 Wastewater Discharge Re-Use in Papermaking Process

- Challenges encountered:
  - Variability of solids level going into the UF
  - Membrane fouling due to organic materials (starches) can result in 1-2 days per week cleaning for "zero-effluent mill"

- New technology needs to increase throughput (flux) and reduce downtime losses:
  - Effective pretreatment chemistries for clarification step; pre-UF (for example enzymes to degrade starch)
  - Effective cleaning chemistries
  - Monitoring techniques to handle variation in incoming stream with proper pro-active treatment response.

12

4 Effluent Monitoring of Waste Treatment Discharge

- The ability to provide reliable quality data for a water stream is critical for re-use. Papermakers do not want to lose efficiency or compromise product quality.

- Some challenges encountered in monitoring water streams exiting a waste treatment plan include:
  - Variability of stream
  - Solids content in the waste stream (challenge to detect COD and TOC)
  - Sensor fouling and long term integrity (scale, microbiological growth)

- Currently on-line sensors are available for oxygen, solids, pH, NH_4, NO_3, PO_4

- New cost-effective technologies are needed for COD, TOC and other species specific to customer concerns.

13

5 Cooling Tower Water

- Cooling tower make up water is one potential area for water re-use.
- The final general concentration guidelines for water quality in the tower can be used for a "treat as needed" concept.

Guidelines for Water Quality IN the Cooling Tower

Parameter	Concentration (ppm)
Phosphate	< 20
Calcium	< 1500
Silica	< 300
Chloride	< 3000
Sulfate	< 3000
NTU	< 50
TOC	< 20

Example of Effluent Quality from WWTP for Coated Paper Mill

	PH	T [°C]	Color CU	COD_{cr} [mg/l]	BOD_5 [mg/l]	SS [mg/l]
Target	6.0-9.0	≤40	≤50	≤80	≤30	≤50
Actual	**7.7**	**35**	**32**	**51**	**7.7**	**22**

14

5 Improvement in Tertiary Treatments for WWTP

- New technologies to improve the quality and reduce the cost of tertiary treatment in WWTP will allow greater water re-use.

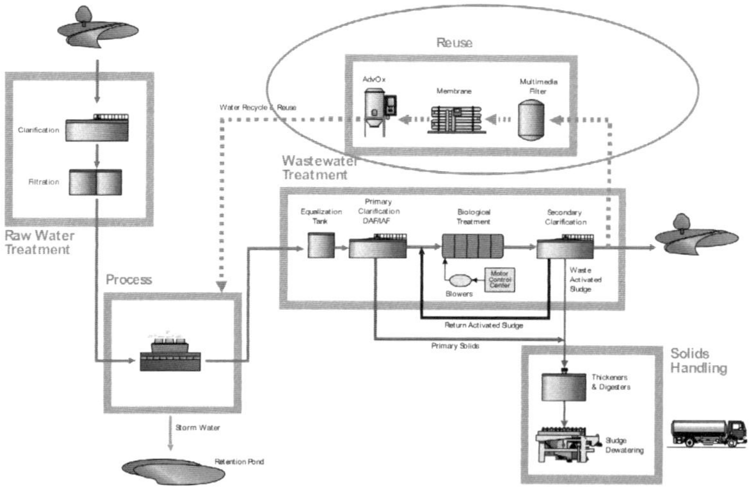

15

Summary

- Innovation is needed in the following areas to promote water sustainability in pulp and paper making:

  1. Pulp washing at higher consistency
  2. Treatment of filtrate from initial Kraft bleaching stages (D_o, E_{op})
  3. Monitoring and control of Save-All clear filtrate quality in order to re-use the water
  4. New treatment and cleaning chemistries to improve throughput of membranes in waste treatment
  5. New technologies for tertiary treatment of waste effluent to supply water with acceptable quality for re-use ("treat to need")

16

WATER
INNOVATIONS
ALLIANCE

Solving the World's Water Problems through Technology

WATER
INNOVATIONS
FOUNDATION

www.waterinnovations.org

WATER
INNOVATIONS
FOUNDATION

Everyone from Presidents to CEOs to farmers to soccer moms say that water is our most precious resource....

....but do they really mean it?

www.waterinnovations.org

WATER
INNOVATIONS
FOUNDATION

At home, the average American uses between 100 and 175 gallons of water a day. That is less than 25 years ago, but it does not include the amount of water used to feed and clothe us.

This illustration show how many gallons of water are needed to produce some meal-sized

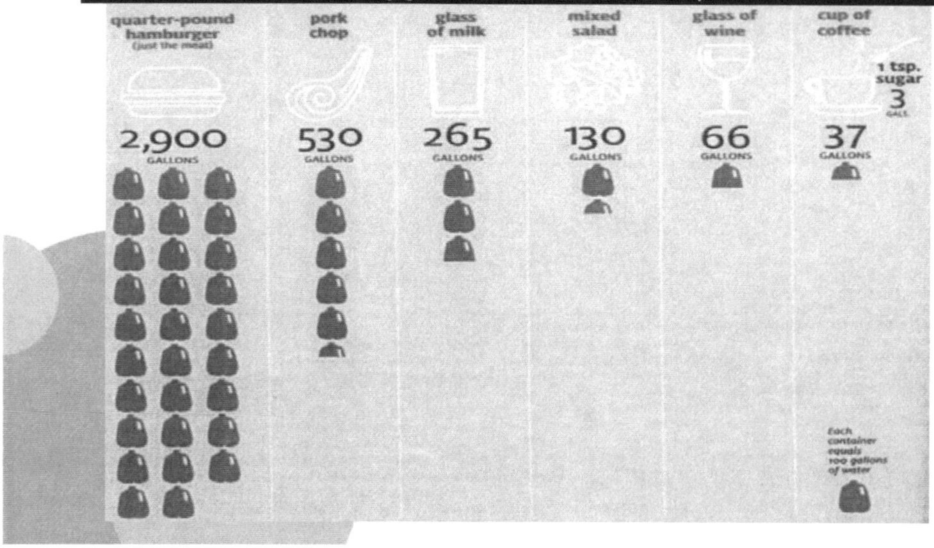

WATER
INNOVATIONS
FOUNDATION

And this is how many gallons of water are needed to produce quantities of some everyday foods:

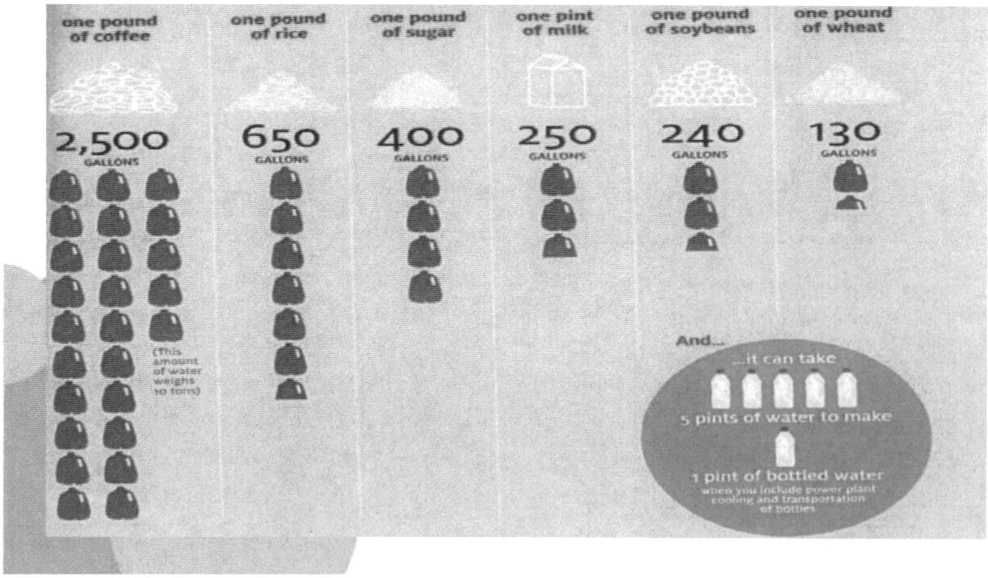

WATER INNOVATIONS FOUNDATION

We have all heard the facts…

Projected Water Scarcity 2025

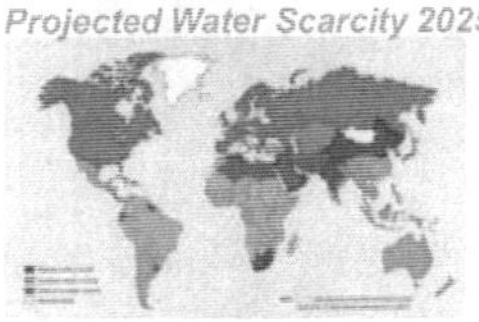

GDP Growth Tied To Water Use

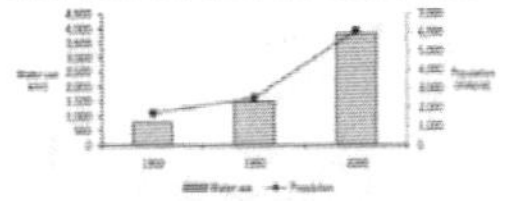

- Only 3 % of the water on earth is fresh water -- little of that is accessible for human use
- Agriculture uses 70% of fresh water supplies
- 1.5 billion people currently do not have adequate access to fresh water …this number is expected to jump to over 3 billion in a decade
- It's the 3rd largest "industry" on Earth--$600billion
- 75% of US urban water infrastructure needs to be replaced. It will take "hundreds of billions of dollars" according to the CIDC
- Developing countries like India, China, and Saudi Arabia need to spend hundreds of billions to put new infrastructure in place
- Growing problems of climate change, energy-water nexus, population growth, pharmaceutical pollutants….

$550 billion this year

www.waterinnovations.org

WATER INNOVATIONS FOUNDATION

Incredibly complex system of overlapping ecosystems, jurisdictions and business interests …

Hydrologic Cycle

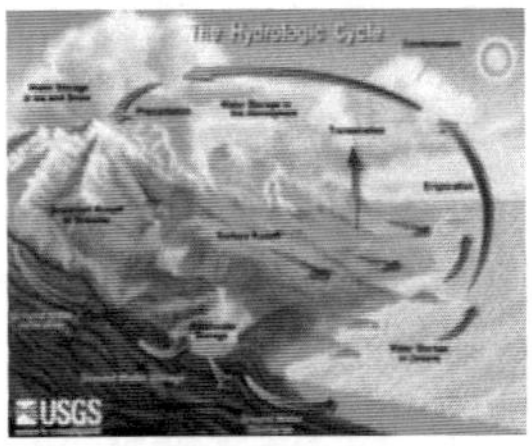

The Water Economy

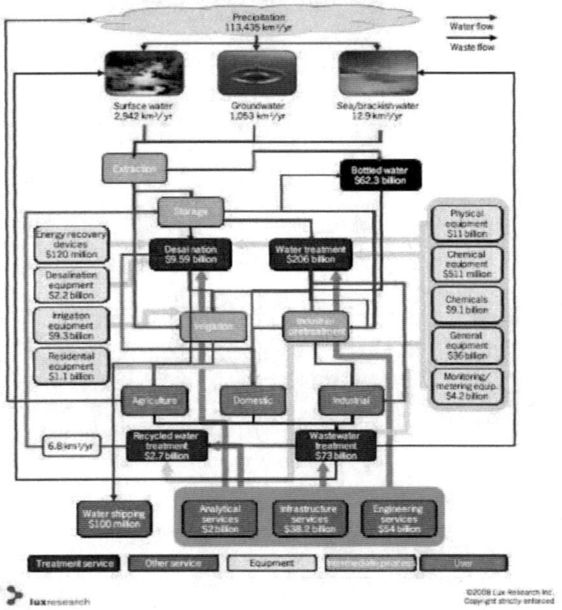

www.waterinnovations.org

OVERVIEW – U.S. WATER SYSTEM

- The U.S. water infrastructure is rapidly aging, decaying and causing disease outbreaks, water loss, and extensive property damage; these problems are primarily caused by ancient water pipes. Many of these pipes have not been upgraded since they were first installed in the years following World War II. Some of these pipes are more than 60 years old and account on an average over 700 water main breaks a day, flooding homes and causing thousands of dollars in property damage.

  In the 2009 Report Card of America's Infrastructure, the American Society of Civil Engineers rated the U.S. water infrastructure at a "D-." The main impediment to making critical maintenance or upgrades is cost. In 1999 the EPA estimated that the total cost of repairing and updating the U.S. water infrastructure was $198 billion – the latest estimate was $335 billion. Recent estimates would set that number at $365 billion.

www.waterinnovations.org

OVERVIEW – U.S. WATER SYSTEM

- Aside from the crumbling infrastructure, we have entered an era of severe fiscal restrictions faced by federal, state, and local governments. Therefore, it is critical that this investment is adequately prioritized and well-directed. The development of a smart water grid system allows for better day-to-day management of a watershed, and the ability to better plan and allocate dwindling fiscal resources. Through the use of innovative technologies for sensing & monitoring, information exchange, and data analytics, a smart water grid system can drive a quantum leap in efficiency of water use and distribution. This would contribute towards maintaining drinking water infrastructure at the traditionally high standards to which the public is accustomed.

  A nationwide smart water grid would create a central command and control center. This would allow for individual communities and watersheds to utilize an advanced water IT system for the management of their water systems. This would create a common standard. Furthermore, this system would allow for data exchange that exceeds traditional community boundaries and looks at where and under what conditions water is accessed, used and discharged. A smart water grid system would have an enormous long and short term impact fiscally, improving human health and the environment.

www.waterinnovations.org

KEY FINDINGS – WATER INNOVATIONS ALLIANCE FOUNDATION RESEARCH AS OF 11-1-11

What is Water IT?

Basically, a "smart water grid. " It's a system for intelligently and effectively managing our interaction with water, via a network of sensors, meters, satellite data, flow monitors, quality testing, and other intelligent devices and management tools.

www.waterinnovations.org

- The Water Innovations Foundation has undertaken research over the past several months, to explore the impact of a water system to better understand its effect on utilities and communities and the potential return on investment. These are some of the initial key findings.

KEY FINDINGS

Why Water IT?

While improvements to physical infrastructure are slow to implement -- faster and lower cost information technology (IT) tools can help manage existing infrastructure more effectively.

www.waterinnovations.org

- \- Approximately 83% of the 53,000 community water systems, serve 3,300 or fewer people.
  \- In 2009, more than 20 percent of water treatment systems in the U.S., which provide water to 49 million people, have violated key provisions of the Safe Drinking Water Act over the previous five years.
  \- The average residential water bill has risen from \$17(drinking) and \$22 (sewer) in 2001 to \$28 (drinking) and \$36 (sewer) in 2010 - approximately a 5.5 percent compound annual increase
  \- Commercial and industrial water bills have risen at even faster rates than residential

KEY FINDINGS

- - It will take 15-20 years of significant investment to stabilize and modernize the U.S. water infrastructure
  - Predicted costs to repair U.S. water infrastructure system: $365 billion
  - Smart Water Grid would create $102 billion in savings (28 percent)
  - Cost to deploy Smart Water network is estimated to be $19.2billion
  - Total Savings: $82.8billion
  - Total Cost to address current water infrastructure problems: $282.2 billion

www.waterinnovations.org

KEY FINDINGS

- - A water IT system will lower annual operating costs on average of 33 percent-- providing more impact on larger systems
  - This is not one project but hundreds of interconnected projects
  - Approximately 15 percent of municipal water is lost to leaks
  - Public water systems that serve more than one million people in the United States average a loss of 17 percent leakage
  - 45 percent of the country's pipes will be rated poor, very poor or beyond repair by 2020
  - Leaking pipes lose an estimated 7 billion gallons of clean drinking water a day
  - A water IT system would allow for significant treated water savings -- nearly 54 percent less leaks for same projected 20 year growth in use and required investment

www.waterinnovations.org

KEY FINDINGS

- - In 2010, Electricity consumption totaled nearly 3,884 billion Kilowatthours (kWh)
  - Electricity use in 2010 was more than 13 times greater than electricity use in 1950
  - The U.S. water system uses over 4 percent of U.S. electricity
  - Water utilities would save 37.5 percent on average if a Water IT system were put in place

www.waterinnovations.org

KEY FINDINGS

- - In addition to lowering energy usage, a water IT system would insure improved environmental conditions, less environmental accidents and reduction of chemical use by a staggering 68 percent
  - Numerous U.S. water experts claim that 55 percent of waterborne illness is caused by improper monitoring, testing and reporting.
  - A water IT system would cut outbreaks of water borne illness by 81 percent
  - A water IT system would lessen the amount of treatment - particularly chemicals - used to ensure safe drinking water by over 18 percent

www.waterinnovations.org

WATER INNOVATIONS FOUNDATION

Examples of water innovations in action...

Veolia Innovation *Accelerator*

QuickTime™ and a TIFF (Uncompressed) decompressor are needed to see this picture.

QuickTime™ and a TIFF (Uncompressed) decompressor are needed to see this picture.

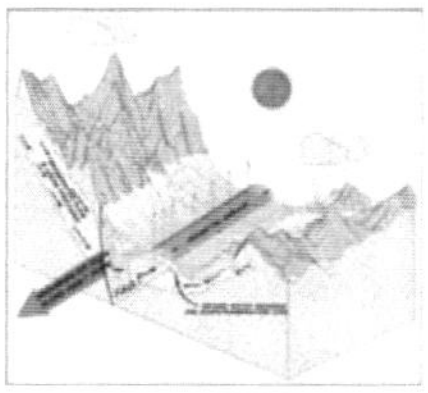

EarthWater Global
MEGAWATERSHED DEVELOPMENT

www.waterinnovations.org

....alas there is just not enough

WATER INNOVATIONS FOUNDATION

Reality Check #1....

Innovation Timeline

- **500 BC** Water Filters
- **300 BC** Aqueducts
- **1854** Chlorine Disinfection

... the way we move, treat and filter water has changed little since the time of Julius Caesar:

- We move water through trenches & tubes
- We force water through tiny holes to clean it
- We put poisons in water to kill other poisonous things

www.waterinnovations.org

Reality Check #2....

....we expect it and accept it:

Federal Spending

- **$1.8bn** Nanotechology
- **$300mm** Federal fleet of hybrid cars
- **$180mm** Volunteers at CNCS
- **$175mm** Water research

www.waterinnovations.org

- Little innovation
- Sporadic technology development
- Institutional and regulatory barriers to implementing new ideas
- Miniscule science and technology funding
- Tweaks instead ***BOLD*** new ideas

Reality Check #3....

... its not getting better:

Water Venture Capital

- Only $130 million invested in 33 deal in '09
- VCs don't get water
- If they did, they almost never do a deal

Water Tech Centers

- Lone NSF water center due to sunset soon
- Less than 10 US university water tech centers
- Budgets rarely surpass $1.5mm annually

www.waterinnovations.org

- Corporations are spending a pittance on mid-to-long term water R&D
- Investors are not funding cutting-edge water technology start-ups
- Only a handful of university water technology centers
- Government doesn't have a coordinated water program and it isn't funding research at an acceptable level

WATER INNOVATIONS FOUNDATION

Why has this happened?

Answer: *The reasons are varied and complex*

"Think of the true cost of water as an unwritten tax on all business and all aspects of your life"

- **Customer Pricing:** Real cost of water is not paid and rarely even appropriately monitored
- **Public Safety:** Water is a key part of human health breeding a "safe & slow" mindset
- **Corporate Atmosphere:** Economic down-turn; M&A's; quarterlies as king; regulatory costs
- **Fix vs. Prevent:** "Fix it" when it breaks mindset
- **Political Backwater//Football:** Water rarely has a seat at the political table and when it does its usually a war
- **Silos:** Tunnel vision on how the entire water ecosystems fits together and managing that bigger picture
- **....and many many more**

www.waterinnovations.org

WATER INNOVATIONS FOUNDATION

So we decided to try and change things. In November 2008, the WIA launched:

WATER INNOVATIONS FOUNDATION

- 501c3 Research Foundation
- Launched Jan. 2010
- Focus on research, reports, standards development and other educational programs

WATER INNOVATIONS ALLIANCE

The Water Innovations Alliance is a 501c6 industry association focused on developing new funding, reducing regulatory barriers, increasing collaboration & raising awareness for cutting-edge water technologies and the problems they solve.

Members includes Fortune 500 companies, start-ups, economic development organizations, universities, NGOs and investment firms.

The Alliance is located in Washington, DC, Shelton, CT & Cambridge, MA

www.waterinnovations.org

WATER INNOVATIONS FOUNDATION

Water Innovations Alliance Members:

MANAGEMENT

Vincent Caprio, COO
vince@waterinnovations.org

F Mark Modzelewski, Ex. Dir.
mark@waterinnovations.org

Advisory Board

- **Alan Hinchman, GE**
- **Carrie Freemen, Intel**
- **Doug McCall, Sensus**
- **Fabien Cousteau, Ocean Explorer**
- **Prakash Chakravarthi, Aqleo**
- **Kelly Carnes, TechVision21**
- **Kevin McGovern, Water Initiative**
- **Daniel Ritter, K&L Gates**
- **Scott Livingston, Livingston Sec.**
- **Delbert Wesley Mullens, Investor**
- **Chris Spain, HydroPoint**
- **Finn Nielsen, Veolia**
- **Josh Wolfe, Lux Capital**
- **Cat Shrier, Watercat Consulting**

www.waterinnovations.org

AUTHOR INDEX